PRIMARY WRECK-DIVING GUIDE

Wreck-diving is a discipline as different from open-water scuba as college is from kindergarten. The basic certification course introduces a non-diver to the subsurface world, and provides him with the essential knowledge and skills necessary to enable him to immerse his body in water and discover that, with artificial devices, he can still breathe. However, a check-out dive and a C-card do not prepare a person for the rigors of wreck-diving.

Most certifying agencies offer specialty courses designed to present the entry level diver with new challenges, encouraging him to improve his proficiency under controlled supervision. This step-by-step approach is good because it ensures that a diver does not get in over his head, so to speak, by taking on more than he can handle. What the beginner sometimes fails to realize is that dives conducted in different environments and under a variety of conditions require an intimate understanding of himself and his limitations, and of the water into which he is about to plunge: the ocean is more than a pool with a larger circumference. By increasing his skills incrementally and by gaining a gradual appreciation for the deep, a diver can achieve his full potential safer and more rapidly.

This book is a primer for one particular and very captivating activity: diving on shipwrecks. It proposes to offer practical information, as opposed to theoretical or mathematical; that is, how to conduct a dive on sunken ships, not what happens to the body under pressure. Furthermore, it intends to address the grimmer realities that are often overlooked: entanglement, equipment flooding, seasickness, and getting lost at sea, to name a few. To a certain extent, these unfortunate events are overemphasized in order to make up for the fact that they are seldom addressed in class or in popular publications: they make uncomfortable enlightenment at best. My intention is not to scare anyone off, but to acquaint people with worst case scenarios that may never occur, and to impart information that is otherwise unavailable.

The topics covered include equipment modification, thermal protection, access to sites, current and surge, wreck orientation and navigation, night diving, photography, pharmacology, and a riveting rivet by rivet account of how shipwrecks got the way they are and why they look the way they do.

From *Century Magazine*, 1886.

PRIMARY
WRECK
DIVING
GUIDE

BY **Gary Gentile**

GARY GENTILE PRODUCTIONS
P.O. Box 57137
Philadelphia, PA 19111
1994

Gary Gentile Productions
P.O. Box 57137
Philadelphia, PA 19111

Additional copies of this book may be purchased from the same address by sending a check or money order in the amount of $20 U.S. for each copy (postage paid).

Picture Credits
All uncredited photographs were taken by the author. The front cover shows the engine of the steam tug *Columbus*, sunk in Gargantua Bay, Lake Superior, Ontario, in 1910. The inset photo is of Tom Le Sage. The back cover photo of the author was taken by Jon Hulburt on the wreck of the *Madiana*, in Bermuda.

The author wishes to thank Ted Green for providing the inspiration to write this book, and for reviewing the manuscript prior to publication, and Barb Lander and Gene Peterson for reading the rough draft; all made valuable suggestions that helped make the finished product better than it would otherwise have been. The author also wishes to thank Ed Hoffman for drawing the illustrations; Gaye Brown, Tom Brown, and Betsy Llewellyn for offering advice on select chapters; and Drew Maser and Pat Stewart for proofreading the galleys.

International Standard Book Number (ISBN) 0-9621453-9-4

First Edition

Printed in Hong Kong

CONTENTS

Introduction

One of the most curious facets of wreck-diving is that few of the rules are inflexible. There is great latitude for personal preference both in technique and in gear arrangement. I am a big believer in adaptability: the capacity to accept input from others, to judge its applicability to certain circumstances, then make appropriate modifications. Rather than tell people what to do, I prefer to make suggestions and provide helpful hints based upon my experience, and leave it to the reader to choose which recommendations to use or ignore depending upon the conditions anticipated on any particular dive. Seldom in any procedure is there a right way and a wrong way; there may be a quicker way, a more economical way, or, in the case of diving wrecks, a safer way. Ultimately, the choice of methodology is yours, and so is the responsibility for that choice.

I also believe in creativity. Each person brings with him into wreck-diving his own talents, skills, and vocational schooling. Thus each diver has something to offer: a fact which should not be overlooked by those whose in-water experience is all that makes them more proficient. Fresh outlooks can often conceptualize new paths through old practices.

While wreck-diving is generally considered to be a leisure time activity, the value of the challenges involved goes far beyond the avocational merit of other hobbies. The training, the experience, and the ongoing confrontation with new and unanticipated situations helps improve a person's resourcefulness: an improvement that carries over into other areas of life. Overcoming the obstacles encountered in diving shipwrecks teaches people to think for themselves, and can help mold a person into a stronger individual by instilling confidence otherwise lacking. It has certainly helped to enrich my own development.

During the course of man's history, tens of thousands of ships have been lost around the world: in the oceans, in rivers, in lakes and other navigable waterways. Throughout the text I habitually use generic terminology based upon the broad assumption that the majority of wreck-diving takes place at sea, and that access is usually gained by boat. Thus the frequent allusion to dive boats and anchor lines. This does not reflect any prejudice on my part, only a recognition that by addressing ocean diving as the norm, I address the mainstream of the activity.

Similarly, one whose entire diving experience has been gained in clear, warm Caribbean waters or on shallow Florida reefs might find many of my references uncalled for. Again, without prejudice, I have put wreck-diving in context with North Atlantic sea conditions, where the water is likely to be cold, dark, and unpredictable: the minimum threshold or lowest common denominator against which most wreck-diving is ranked. Those who have not yet experienced the difficulties of adverse environmental conditions should keep in mind that this is what sets the tone of the book.

Chapter 13, "Room for Ladies," is about women in wreck-diving. While the major questions raised involve gender specific issues, I intend the information to be carefully considered by men. Only through a greater understanding of the difficulties encountered by women can men help achieve the harmonious and homogeneous social structure that must dominate twenty-first century thinking. Hopefully, a change in behavior will lead to a change in attitude. And, while change is always painful, it is also necessary for longterm cultural survival.

Ponder these thoughts while you pack your gear. Eat, drink, and be merry, for tomorrow we dive.

Anatomy of a Shipwreck

If you picture a shipwreck simply as a ship that happens to be submerged, you couldn't be further from the truth. To the uninitiated, this fanciful imagery comes from spending one's formative and impressionable years watching Saturday morning cartoons, afternoon matinees, and unsophisticated Hollywood productions; or by reading submarine adventure novels written by authors whose knowledge of the sea is largely vicarious. They depict shipwrecks as upright, undamaged vessels whose decks are clean, whose sails are set, whose cabins and corridors are uncluttered, and whose intact hulls rise sheer off a white sandy bottom; the water is always wonderfully warm, and visibility is never less than perfect.

In the real, underwater world, however, a shipwreck seldom looks like a ship; it is not a pristine vessel of wood or steel placed gently in a preserving bath, to lie for all eternity unharmed and unchanged. Although it may seem too obvious a statement, a shipwreck is a ship that was *wrecked*. Whether it be by storm, stranding, collision, foundering, or the wicked wages of war, a ship once wrecked is a ship no more; it is only a caricature of its former self: distorted, twisted out of shape, and forever altered from the fine lines that long ago slid down the tallowed ways. Moreover, once condemned to the racking ravages of nature, and absent constant maintenance, the normal degenerative process is accelerated. Wood rots and steel rusts at an ever quickening pace until eventually nothing remains that is recognizable as the craft of man.

Somewhere between the majesty of architectural grace and the dissolution to atomic particles lies that nebulous quantum that we call a shipwreck: a stage in the progression from a highly organized structure to abstract chaos. Viewed from widely separated points in time, a shipwreck may seem to have undergone a sudden and dramatic transformation, especially when the present-day wreck is compared to a photograph taken during the ship's historic launching. Dramatic that transformation certainly is, but sudden it is not.

A comparison might be drawn between the caterpillar and the butterfly: two stages of insect metamorphosis that are visible to an observer. Inside the tightly woven crysalis a biological change takes place, slow but steady and preordained. Several weeks or months after the furry, multilegged creature encases itself in silk, a winged, streamlined flyer emerges. Yet in between the larva and the adult is a pupal stage during which the unseen change occurs. If we could part the delicate strands of the cocoon or somehow make it transparent, we could watch the day-to-day development from one form to another.

Similarly, a shipwreck is not a static singularity but a process at work: in this case a dynamic system of degradation cloaked in an aqueous solution which is not an embalming fluid that protects and preserves, but rather a slow-acting acid that eats away relentlessly at everything it touches. Shipwrecks are not immortal; they are way stations to oblivion.

For more than two decades I have watched my favorite wrecks disintegrate. From one season to the next they crumble, sometimes radically, especially after a winter of heavy storms. These changes were not always undesirable from an explorer's point of view, for sometimes previously unreachable areas became readily accessible when hull plates fell off and offered new entry points to the vast interior. Usually, though, as wrecks collapse they look less like a ship, become less comprehensible, and are less interesting to explore.

Whenever I dive a wreck for the first time, I understand that I am seeing but a moment in its aging deathspan, that I have sliced open the watery cocoon at one particular instant in order to view the metamorphosing structure within. All the rest of the shipwreck's history is left to the imagination. But with the proper knowledge of a ship's construction and the forces causing its decay, not only can I envision how the wreck arrived at its present state of collapse, I can extrapolate mentally how it will appear in the future: an extra reward for the inquiring mind.

Before we can begin to learn how to dive a shipwreck, we need to understand what a shipwreck is and how it got to be the way we find it: what forces act upon it, what processes are involved in its destruction. In this chapter I will begin with some broad generalizations by describing the "ideal" shipwreck—one that you will never see. Then I will portray the various permutations of collapse that might occur depending upon the type of ship, the

Two fanciful versions of windjammers wrecked at sea. (From *Twenty Thousand Leagues Under the Sea*, circa 1872.)

methods and materials of fabrication, the circumstances of loss, and local environmental conditions that contribute to the speed of ruination. From there it is up to you to assimilate the different scenarios and relate them to a particular wreck with which you are familiar. It should prove enlightening.

My *American Heritage Dictionary* defines "ship" as "any vessel of considerable size adapted for deep-water navigation," and as "a vessel intended for marine transportation, without regard to form, rig, or means of propulsion." Roget's College *Thesaurus* gives one hundred twenty-five examples of which about half are the smaller variety called "boats." Generic headings that commonly come to mind are tankers, freighters, liners, tugs, and barges; the various sail configurations of windjammers (brigs, brigantines, barks, barkentines, schooners, etc.); and warships (battleships, cruisers, destroyers, gunboats, and submarines, to name a few modern types, and men-of-war, frigates, and ironclads of the older class). The remains of all of them can be found at the bottom of the sea.

Even a ship that floods slowly and slides inexorably beneath the waves will sustain some form of damage in passage. Sinking a ship in a hundred feet of water is not the same as dropping a Volkswagen from the top of a ten story building, but there is bound to be a shock when the keel touches the sand. Nevertheless, let's suppose that the ship was not seriously injured but landed lightly on the bottom with nothing to show for it but a good soaking.

That ships are designed to operate *on* the water rather than under it is significant and often overlooked. When the hull is flooded, the external pressure (exerted by the water against the ship's sides) is removed; without that support, the strength of the steel plates is weakened since the design criteria take that constant support into account (the same as a bridge builder figuring on the solidity of the rock on which abutments are placed). In other words, a ship's hull is constructed primarily to keep the weight of water out, not the weight of cargo in: a delicate balance that is upset immediately upon submergence.

Thus a heavily laden cargo vessel may be unnaturally stressed before it ever reaches the bottom, because the cargo pushing out against the hull is no longer counteracted by the water pressing in. This uneven stress continues to exert an outward force after the vessel comes to rest, and eventually the hull is pushed apart from the inside like a chick hatching out of an egg. As the wood or steel hull is pried outward, the decks are pulled out from their attachment points and fall down into the middle of the wreck. Thus an upright wreck can "accordion" in stages until it flattens completely; the wreckage will appear extremely broad because the hull plates that once rose tens of feet upward are spread out to both sides.

This process of collapse may take decades. If the wreck lies in shallow water more prone to subsurface wave action generated by atmospheric storms, it might take only years. If the wreck grounds on a shoal or near the beach, where part of it lies awash or just under the surface, it might be torn apart in days or weeks, or during the next gale.

Even without the outward push of cargo to accelerate the process, any upright wreck will eventually break down in the same manner. If the hull is wood, teredoes (wood-boring mollusks) will eat through the organic substrate until the beams are riddled with holes and lose their structural integrity. If the hull is iron or steel, sea water will oxidize the metal until it no longer has the strength to support itself; rivets will rust away, welded seams will crack. Electrolysis will strip electrons from exposed metal surfaces through galvanic action, and cause pitting and fragility.

In fresh water the time span of collapse is greatly extended, perhaps to the order of centuries—because teredoes live only in the marine environment, because fresh water inhibits rusting and electrolysis, because the constant cold retards biological degeneration and metallurgical destruction. In my experience, the most intact and fascinating shipwrecks are those that grace the bottom lands of the Great Lakes: a veritable museum of lost ships of all types and ages, and a wreck diver's paradise. But nothing can prevent the inevitable; it can only be prolonged. Local divers have gone as far as to emplace steel rods across the upper strakes of some wrecks, in order to forestall the outward spread of the hull. But this situation is the exception rather than the rule.

Our "ideal" shipwreck is bound for ultimate annihilation. The only portions that may be saved for near-time posterity are those parts that soon become covered with mud, silt, sand, or other loose sediment. Just as your feet are sucked into the sand on the beach when waves wash over your ankles, so does a shipwreck work its way into the sea floor: pulled down by its weight into a supporting sediment with less specific gravity. After a while the remains of a wreck may be totally buried, visible only to electronic sensors that can "see" through the bottom.

Hardhat divers salvaging a paddlewheel steamer. (*From Century Magazine.*)

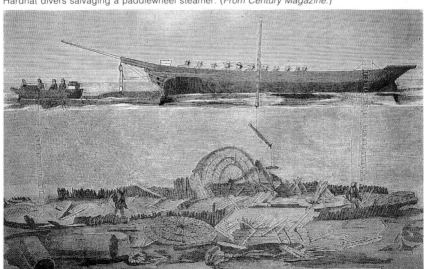

Most of the broken down wrecks we dive are largely embedded but have discrete portions exposed, rather like mountain peaks poking through the cumulus while the lower elevations of the range are obscured by cloud cover. A typical example is the wooden windjammer commonly called a "snag" because so little of it has survived the onslaught of microbial action. We see three roughly parallel upright rows of timber that merge toward one end like the optical illusion of railroad tracks coming together in the distance. These timbers are the vertical extensions of the bottom of the hull which, lying horizontal, is usually covered over. The center row is the keelson (the main structural member that lies above the keel and that runs longitudinally the length of the ship); the two other rows delineate the outer edges of the hull where it angles upward to form the sides (called the "turn of the bilge").

Depending upon the level of the sand, the hull may be completely buried, partially uncovered, or largely exposed. These conditions of exposure can change slowly throughout the years or dramatically after a big storm. Sand can alternately build up on a wreck (called the "depositional phase") or it can wash away (called the "erosional phase"). The amount of exposure may vary on a random basis or it may be cyclical (such as when the Gulf Stream moves close to shore during the summer and away from the shore in winter).

One reason that experienced divers visit a relatively flat wreck site again and again is because they know that the appearance of a wreck changes continually. Lobster holes open and close, debris fields come and go, deep

Ed Hoffman's illustrations on this page and the next capture the evolution of a sunken windjammer's degradation.

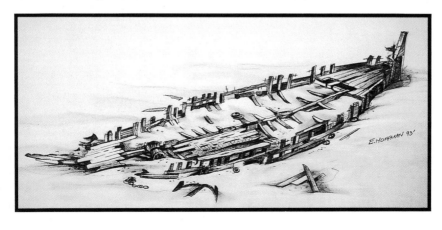

14

washouts occur. Artifacts once interred may become partially or completely exposed by the removal of a thin layer of sand; or hidden structural components may reveal themselves to the astute nautical enthusiast. Thus a previously boring and uninspiring wreck may suddenly yield a cache of concealed excitement.

Another typical example is the steel-hulled steamship. It goes through stages of collapse similar to the wooden windjammer, with the hull peeling outward and the decks landing approximately along the center line of the wreck. The steel plates may fall off one at a time, in which case they are likely to knife straight down into the bottom sediment close to the turn of the bilge, then fall onto their sides and become quickly buried; for a while this will leave the wreck with a "skeletal" look because the stanchions and cross beams may still be in their molded positions; later, the framework will fall away and land farther from the main structure of the wreck than the hull plates that fell off first. Alternatively, the sides may bend outward with the steel plates still attached. Or, some plates may fall off and others remain in place until the whole bulkhead goes, resulting in a jumbled pile of twisted beams and bent hull plates.

It is not uncommon for the midship bulkheads to collapse while the bow and stern sections remain intact. This is due to the greater strength provided by the narrow beam at the ends, and to reinforcement of those portions of the ship for cleaving the sea and for supporting the propeller shaft and rudder.

Ed Hoffman envisions the process of deterioration of an iron- or steel-hulled steamship.

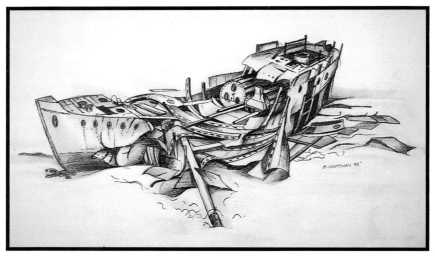

Thus different sections of a wreck may appear in various stages of collapse, all at the same time. A typical configuration for a steamship is one with the bow intact or partially broken down, a long area of low debris extending aft to the midship superstructure (where the height of wreckage may increase slightly), exposed boilers and engine, and an intact or partially broken down stern. Eventually, the bow and stern will collapse and leave the boilers and engine as the only highpoints. Usually, the engine will have the highest relief, sitting like a tall monument in the middle of a junk yard. Around the engine and the boilers you are likely to find portholes and other long-lasting items that dropped from the superstructure when it rusted away.

Current can affect the condition of a wreck. A prevailing current (one that commonly travels in the same direction) will deposit sand on the weather side and erode sand from the lee side. If the current is extremely strong it will gouge deep scours next to the wreck and leave long, down-current patterns. I have seen washouts and siphon holes over ten feet deep, looking like bomb craters or large inverted anthills. If the wreck is tall and intact, the lee side may be so well protected from the prevailing current that deposited silt accumulates to a considerable depth, and covers any fallen debris.

A prevailing current may also have localized effects on a wreck's appearance and state of preservation. The weather side may have a substantial population of filter feeders such as coral, hydroids, and sea anemones, while the lee side is relatively devoid of these animals. This is because the microorganisms on which filter feeders survive and grow are carried by the current, and where there is less current there are less nutrients to support a large population. This effect is sometimes apparent on opposite sides of a vertical steel plate: the side facing the current is densely carpeted with waving tentacles while the side protected from the current is practically barren, and may be inhabited by only a few small creatures whose nutritional needs are easily met. Thick marine encrustation makes it difficult to discern shiplike shapes and contours: you might look directly at an unbroken porthole without recognizing it.

The porthole on the left is obvious, but the one on the right (with the glass intact and thickly encrusted) may escape the casual observer.

 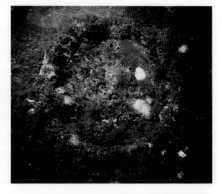

In the photo at left, taken on a wreck in the cold fresh water of Lake Superior, ribbons of oxidized iron drool past a bronze porthole. In the photo at right, a diver hammers on a porthole on a wreck in Bermuda, where sea fans and giant coral heads nearly obscure the contours of the wreck's hull plates, making the wreck appear more like a natural reef than an intrusive man-made object.

A wreck's features can also be obscured by calcareous growth: calcium carbonate that precipitates out of the water and forms a variety of bulbous configurations. The process is similar to the growth of stalactites in limestone caves, except that in the latter case gravity tends to elongate the resultant accumulation of precipitated material because it forms overhead from calcium seeping through the ground. Calcareous growth is a chemical process; it is not organic.

When not overburdened by the presence of marine encrustation, iron will sometimes flow as its surface oxidizes and produces particles of rust. This seeming fluidity is actually the molecular migration of iron oxide in a process similar to that of the mineral accumulation that stimulates crystal formation. Rust can form long slender ribbons that resemble fat bacon strips with one edge attached to the surface. If this ribbon "pours" off a projecting lip the iron oxide can take the shape of an icicle; then it is call a "rusticle."

A cleansing effect can result from a fast predominant current, which transports particulate matter at such high speed that exposed surfaces are sandblasted to bare metal. Brass might have a polished sheen, while iron or steel might show a reddish tint from increased oxidation. Rust is not always evident because the microscopic flakes are carried away as they form. In this case the metal keeps getting thinner until holes appear in the flat surface; then you can observe how flimsy the steel plate has become.

It is generally acceded that the structure of a wreck, whether wood or steel, is protected and held together longer if covered by thick layers of encrustation, the rationale being that the organic overlay acts as a barrier that wards off teredoes, and reduces oxygenation and electrolysis. Coral also acts as a cement, as anyone who has ever tried to pry loose an artifact in tropical waters can attest. Nevertheless, teredoes, oxygen, and free ions will eventually find their way through the porous coating and freely interact with the wood and metals of which ships are constructed. Finally, all that remains are items made from materials that are more resistant to corrosion: porcelain, rubber, brass, copper, silver, gold, etc.

Even our "ideal" wreck must eventually collapse, not in a pure arithmetic progression like a dissolving cube of ice, but in a series of staccato starts and stops like a complicated ice sculpture whose castle ramparts topple over as its foundations melt and weaken. The chances are against a diver actually observing a deck caving in or a hull plate tumbling over, for these gross acts of demolition are more likely to occur when a sudden extraordinary force is applied: a severe atmospheric disturbance whose furious wrathful fingers reach deep beneath the waves in wanton abandon. When wave heights climb into the double digits, dive boats are apt to be tied securely to their docks.

When Hurricane Andrew obliterated Miami in September 1992, Florida's artificial reef program was arbitrarily reorganized. Recently scuttled reef wrecks were torn asunder and flattened, some that lay as deep as two hundred feet. One intact wreck was rolled over on its side, another was moved several hundred yards. Of course, an historic wreck that is firmly embedded in the bottom sediment is not going to move, but it can be severely thrashed by deep wave action. If you've ever watched a documentary program showing a beach resort being trounced by incoming breakers and houses being swept off their pilings, you can understand the wallop that a wall of water has. The fact that a wreck is already in the water does not significantly lessen the blow.

Storm generated wave action can not only knock a wreck apart, it can scatter wreckage over a wide area. Smaller items without great weight (china, bottles, and the like) or with low specific gravity (wooden beams and planking) may be carried down-current so far from the original site that they are lost forever; or, if found, there is no way to establish a relationship between the items and a specific site, especially near inlets, harbors, and dangerous shoals and rocks pinnacles where many wrecks are likely to be concentrated.

There is a classification of wrecks called "buoyant wrecks": the remains of wooden hulks that are buried on the beach or just offshore, and which are repeatedly "refloated" when a storm unearths them. The current carries them off, sometimes for miles, until they run aground on another shoal or headland. Usually these are not complete wrecks, but hull segments held together by planks and spikes.

Working in concert with current is surge: the reversing motion of water that is often violent, particularly in the near-shore shallows. You know what happens if you keep bending a thin sheet of metal back and forth: it snaps in two. The same thing happens to hull plates after years of buffeting in the surge: another reason why tall stately ships end up looking like low heaps of scrap metal. No thickness of steel or amount of encrustation can fail to yield to continuous surge.

The interaction of all these destructive forces and processes paints a complicated picture that only accounts in part for the way we are apt to find a shipwreck lying on the bottom. Under less ideal circumstances a wreck may break up more quickly, may be disbursed more widely, and may collapse upon itself in such a way that previously segregated components will be found juxtaposed nonsensically. Let's look at the different modes of sinking, and see how they affect a wreck's initial placement and subsequent state and rate of collapse.

World War Two hydrophone operators often reported hearing "breaking up sounds" as torpedoed enemy vessels sank in deep water. The awful creaks, groans, and hisses that they heard through their headphones were the bending of inner bulkheads as compartments were crushed by the influx of water, the rending of decks as they were pried apart by the suddenly compressed air, boilers bursting when icy water touched their red-hot casings, ruptured steam lines, water gushing through the ship, and air gurgling out through ports, hatches, and forced-open seams. The sinking of a ship is a traumatic event.

Added to this mass interior destruction is a vessel's attitude as it takes the final plunge. A ship does not always sink upright on an even keel: it may go down either bow or stern first, it may roll onto either side, it may capsize, it may list at any angle of inclination between straight up and upside down, or it may head for the bottom and come to rest in a combination of attitudes between fore and aft and side to side; for example, it may roll onto its port side then sink by the bow.

Knowing the circumstances of sinking may help you understand the layout of the wreckage and tell you where to look for worthwhile relics. It's important to know that a ship took a 45° list to starboard before it disappeared beneath the waves, because it means that the loose paraphernalia (furniture, supplies, navigational equipment, personal belongings, etc.) slid to the starboard side, and is likely to have come to rest against the starboard bulkhead, or against the port side of vertical uprights in the interior compartments—even if the ship righted itself after hitting the bottom, because the stuff cannot slide back to its original position. If the wreck is intact, artifacts are likely to be found on the starboard side; the same principle holds true after the wreck is completely broken down and its sides have disintegrated.

A good example of how a steel-hulled ship "thins out" as its bulkheads deteriorate. At the perimeter of the rust holes the steel is knife-edge sharp; not only will these metal edges cut bare flesh and exposure suit material, but they will act as "grabs" on goody bags and loose equipment.

In this case, too, the remains of the upper hull and superstructure will be found on the starboard side of the wreck. Not only will fittings from the port hull be interspersed with fittings from the starboard hull, but rubble from the highest decks may be found away from the wreck and no longer contiguous with the main debris field. This latter situation arises because ships are very tall in relation to their width; as the superstructure and upper decks deteriorate and slide off into the sand, a separate debris field is created. Years later, after the wreck has wasted away to an indecipherable pile of rubble that is partially sanded in, the top of the bridgeworks may protrude from the bottom some thirty or forty feet off the main structure. If the visibility in the area averages only ten or twenty feet, you may not know that another part of the wreck exists just beyond the limit of vision.

The converse corollary of the scenario described above is that by observing the morphology of a wreck site and the arrangement of debris, you can determine the circumstances of sinking: a useful bit of information that may help you to better appreciate the wreck and its condition, and that might aid in identification if the name is unknown.

The Poseidon Adventure notwithstanding, ships capsize only on rare occasions. The primary reason is ballast. Wooden windjammers had their bilges filled with sand, gravel rock, or pig iron; modern vessels have deep tank compartments full of water. In either case, weight close to the keel gives a ship stability and helps keep it upright the same as a packet of sand in the base of an inflatable schmo doll.

Heavily armed warships are the most likely ships to capsize. This is because of the incredible weight of the guns and turrets, which for effectiveness must be positioned high on the upper decks. Warships are properly ballasted to prevent them from overturning under normal circumstances, but when that balance is upset by the sudden and massive inflow of water due to explosive damage below the water line, they quickly lose their stability. If counterflooding cannot restore equilibrium, they will roll over completely and, once on their beam ends, will keep right on going, "turning turtle" as the saying goes, because a new equilibrium is established with the guns and turrets becoming the bottom center of gravity. The distribution of armor, predominantly above the water line, also contributes to the disparity between

Left: the point of the bow with a "skeletal" look due to the sloughing off of hull plates. Right: an inverted wreck with the keel uppermost.

Left: With 100-foot visibility one can see from the engine (in the lower left corner) clear to the upright rudder. All else is low-lying debris. Upper right: A diver explores a field of disarticulated hull plates. Lower right: A diver studies the wooden planking of a fresh water wreck.

ballast and topside load. For comparison, think of a warship as a vertical dumbbell, with heavy weights at the top and bottom.

Tankers have somewhat of a tendency to flip over after suffering severe bottom damage that results in uneven flooding, because of the way the tanks trap air. An ocean liner whose compartmentalization has been broached, such as by fire, may turn bottom up due to the weight of the many passenger decks stacked atop each other. A wooden sailing ship that rolls over on its side may invert completely as its cargo and ballast pour out; then it may drift along with the current for weeks or even months, sometimes traveling thousands of miles. At one time this was so common that special derelict destroyers were commissioned to seek out and sink these floating hazards to navigation.

Generally speaking, a capsized wreck maintains its hull shape longer than a wreck sitting upright or lying on its side. This is because the forces against the hull are directed inward, giving it the same kind of strength as an egg but reinforced with steel or thick oaken beams. Not for many years do hull plates loosen enough from their fastenings to let go; and then, they often cannot fall away because they are resting against the support beams which are themselves mutually supportive.

Barring other circumstances, such as massive staving in of the forward compartments, steamships generally go down stern first due to the weight of the machinery situated aft. If you've seen wartime newsreel footage of steamers sinking after a torpedo hit, with the bow rising high in the air, you know what I mean. If the ship goes vertical, measuring its length before knifing backward into the depths, you can forget about diving it because the depth must be more than the length of the ship.

What you should note in those films is the foaming fount of water that continues long after the ship is gone. As a ship goes down by the head or stern, water cascades into every orifice as soon as it submerges. The deeper the ship sinks, the faster the inflow. Soon there is a raging river of water rushing through the interior, crashing through doors, knocking down partitions, sweeping everything before it that is not bolted down, and creating general havoc. Trapped air is compressed until it finds a way out, often by popping the glass out of the ports. A wreck may leak air for several days or weeks. As the pressure of trapped air is released, the buoyancy it offered is lost and the ship settles down even more, pressing against structural components that were never intended to support the weight of the hull

Ships were not designed to lie on their sides. Even under the best conditions, a ship that alights on its side will soon begin to fall apart. Torsion will sheer off bolts and rivets and cause welded seams to crack. Guy wires will break; booms, funnels, and ventilators will topple. Deck cabins and superstructure will slough away. Interior partitions and stairwells will tumble into a disjointed heap, crushing everything in the way. Once the lower decks rust or rot through, and the overhead hull loses its support, the wreck will slowly crumble like a melted plastic model in slow motion.

In the old days trapped air had a more visibly dramatic effect: since many ships had wooden decks (even steel hulled ships), these decks were sometimes lifted right off the ship as the hull sank out of sight. On occasion, survivors clung to the floating decks and associated debris, and were carried away from the site of the sinking. It is important for the wreck explorer to know this because it may explain the absence of ship's appurtenances you would ordinarily expect to find on a wreck. There are accounts of wheelhouses tearing free and remaining afloat for days.

A ship that grounds on an offshore shoal or that comes to grief in the surf during a storm is likely to be battered to pieces by the waves, sometimes within hours. There are many cases on record of wooden hulled ships splitting apart and being scattered over several miles of beach front. Often, survivors have come ashore on floating deck houses and sections of decking. Such shallow water wrecks are at the mercy of the seas, year after year. There may not be much left of them today except embedded hulls and heavy metal parts. Debris can be widely scattered. The ambitious wreck diver will explore away from the main hull in order to look for related materials.

There are also many post-sinking factors that aggravate a wreck's inclination to deteriorate. A ship sunk in the shallows of a northern bay or lake may be subject to excessive ice damage. Ice forming around the structure of a wreck protruding above the surface is not by itself very destructive; but when

Left: A wooden-hulled ship being pounded by waves on the beach. (Courtesy of the Outer Banks History Center.) Right: Ice moving with the inexorable strength of the tide grates against a hull made of ferroconcrete: the steamship *Atlantus*. (See pages 58–59 for more about the *Atlantus*.)

the ice moves with the current or tide, or is propelled by the force of the wind, it has the power to cut into metal or wood as effectively as the best industrial grinder or chain saw. Crystalline blades can cut several feet below the surface, and what the ice does not cut it can pull apart, perhaps damaging a wreck down to its very keel.

People have a tendency to underestimate the immense size of ships. The average steamship or windjammer sitting upright on the bottom at a depth of a hundred feet is likely to have mastheads that reach the surface. Survivors often clung to the masts until they were rescued by a passing vessel. Later, the same masts that saved human lives became a menace to navigation, and had to be removed. The job was sometimes done by divers who laid explosive charges on the deck where the masts came through, delivering a clean cut. But more often than not, timed satchel charges were lowered to a wreck from a salvage vessel: a method that was just as effective but a lot more destructive. No one cared at the time, since wrecks did not become culturally significant until twentieth century bureaucratic fiat made them so.

Some ships sunk in shallow shipping lanes and harbor approaches had to have their structure lowered in order to prevent deep-draft vessels from ramming the sunken hulks or scraping across them. Many a wreck has been virtually obliterated by several tons of dynamite in order to ensure the safety of latter-day shipping. At the very least, a wreck might be wire-dragged to cut down stacks, masts, and superstructure. In wire-dragging, a steel wire several inches across is stretched between two vessels steering a parallel course, with the wire suspended at a predetermined depth; the wire cuts through a wreck's upperworks like a taut piano wire through butter.

Salvage operations may account for degradation of a wreck that seems excessive for its age and environmental position. In order to reach a ship's cargo, commercial salvors usually blow out the sides of the hull, or blast the decks to smithereens. This kind of gross destruction is in keeping with the fact that a wreck is, after all, just a wreck—with no value other than what can be recovered from it.

Many ocean wrecks that were not commercially salvaged or demolished as hazards to navigation owe their advanced state of collapse to the exigencies of war—depth charge attacks. As German U-boats preyed mercilessly on innocent merchant ships during World War Two, the Allies fought back by dropping underwater explosives set to detonate at predetermined depths. Sonar (sound navigation ranging), a device that transmits an acoustic signal that is reflected off a submerged object and detected by a receiver, could not distinguish an enemy submarine from a sunken ship. Consequently, depth charges rained down upon more than a few shipwrecks and caused considerable damage: not only by flattening a wreck's structure, but by fracturing hull plates and exposing more surface area to the sea, thus increasing the rate of deterioration. As the war progressed, the technology became more sophisticated and sailors became more proficient; accuracy increased dramatically. Furthermore, there was a belief at the time that German U-boats "hid" alongside shipwrecks in order to disguise themselves from penetrating sonar waves; standard operating procedure for Naval escorts was to drop harassing depth charges on any suspected target in order to thwart this suspected practice. This is another reason why we may find the profiles of large tankers and freighters razed and ruined and more devastated than can be explained by the forces of nature and time.

In summation, in-water experience is the best way—perhaps the only way—to gain insight into the progression of a sunken ship's collapse. After seeing and exploring many wrecks of all types and ages, you will develop a "feel" for the processes at work, you will appreciate the inescapable certainty of a wreck's ultimate demise, and you will understand fully that nothing man has ever built will survive forever the relentless onslaught of nature. Wrecks change from year to year, and from storm to storm; ultimately they will disappear in the vast wasteland of time. But for now, at least, they are here to explore.

All that is left of this composite-hulled vessel are the iron ribs and brass spikes that once held the wooden planks in place.

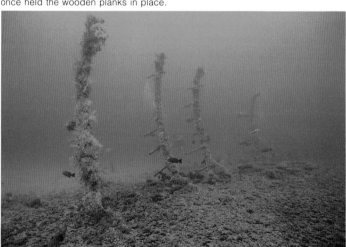

Gearing Up for Going Down

Wreck diving is an equipment intensive activity, perhaps more than any activity other than walking on the moon. Aside from the obvious need to carry a sufficient quantity of atmosphere to permit you to stay on the bottom for a reasonable length of time, and a reliable means of trickling that atmosphere into your lungs as needed, a shipwreck creates an environment in which it behooves you to anticipate having to deal with circumstances beyond your ordinary experience. By this I don't mean to imply that a shipwreck is inherently hostile (it is, after all, only an inanimate object), but that in order to enjoy and fully appreciate it you should outfit yourself with all the gear, tools, and paraphernalia you are likely to need in order to explore the wreck safely, and that will help you to achieve the objectives of the dive.

Some divers take this adage too far, perhaps because they were scouts in their youth, and go in the water prepared for every possible contingency. They look like refugees carrying all their earthly possessions, or like hardware store stock boys. There's no need to dive with everything you own, only with what you honestly expect to use: you can always go back another day with different tools for different jobs.

The purpose of this chapter is to present helpful hints and provide little pointers that will make your dives go more smoothly, while enabling you to accomplish more on a dive without extending your bottom time: in short, to increase your efficiency under water. In addition to modifying basic open-water scuba configurations for wreck diving use, it is equally important to wear accessory items or carry them so they are out of the way yet readily accessible. At first reading, these adaptations may seem complicated. But by taking a step-by-step approach, and gradually incorporating wreck-specific techniques into your diving regimen, gearing up for a wreck-dive will soon become second nature.

Presenting your best profile may seem like advice more appropriate to portrait photography than wreck-diving, yet I must begin by stating categorically that *the* most important feature of fitting out to dive on wrecks is "streamlining." Wrecks are notorious for snagging the unwary diver or loosely hanging equipment.

Large mesh bags clipped to weight belts or left dangling on long lanyards are begging for trouble. These bags are necessary for wreck divers who collect things: fish, shells, and souvenirs, for example. They are usually called "goody" bags; in areas where lobsters predominate, they may be called "bug" bags because lobsters are referred to as "bugs." Mesh bags also serve to carry tools and extra gear on a dive. When not in use, the mesh bag should be folded into itself in order to make it as small as possible. Otherwise it will catch on wooden projections, minute metal spurs, and coral.

The diver on the left is asking for trouble—his goody bag is hanging down way too far, and when he swims horizontal it will swoop across the wreck like a trawl net. The diver on the right has his goody bag tucked close to his body.

Mesh bags come in different sizes. Use a small one for carrying tools, liftbags, or line reels; carry another for goodies. An empty bag has hardly any weight in the water, so if you use a wrist lanyard the rope may float off your hand unnoticed, especially if you are wearing thick neoprene gloves. An elastic lanyard can solve this problem, but it may break under the strain of a heavy artifact or big lobster, especially when climbing a boat ladder, when the weight is no longer buoyed by the water. I have my bags rigged with three means of attachment: a rubber inner tube strip or a loop of surgical tubing, a nylon lanyard, and a brass snap hook for attaching to D-rings; then I can carry it or clip it off as desired. When I carry the bag by hand, I loop both the elastic and nylon lanyards over my wrist.

I do not recommend clipping anything to a weight belt unless it is inexpensive and easily replaced. The reason is that weight belts are designed to be ditched in an emergency, in which case anything attached to it is lost. Worse, too many times divers have been unwilling to drop their weight belts because of the monetary investment. This has resulted in more than a few fatalities. Because of this, the more progressive dive clubs have instituted a policy of reimbursing divers for the cost of replacement of a ditched weight belt, the rationale being that the lack of pecuniary loss will inculcate a frame of mind in which a diver will more readily let go of the lead before it is too late to save his life. If you want to clip off your gear, clip it to a utility D-ring on your backpack belt or buoyancy compensator.

Whenever a snap hook is used to clip gear to your person, the snap hook must be attached to the gear by a piece of line called a leader. That way, if the gear becomes hopelessly entangled in wreckage or is fouled in such a way that you cannot see how to untangle it (due to darkness or to the limited field of vision provided by your mask), the leader can be quickly severed with a knife. Do not be above pulling off your gloves despite the frigidity of the water, in order to increase dexterity and to be better able to feel the entanglement.

Perhaps the most overlooked cause of snagging is catching the regulator hose on a projection. Having the second stage ripped out of your mouth on an inhale is not a pleasant experience, and can easily lead to panic, justifiably so. This situation can be ameliorated by not being "rammy" on the wreck: swim with slow, easy motions and practice full awareness of your surroundings. After all, this is exploration, not a downhill slalom.

You can also reduce the possibility of such a snag by shortening the hose. Too many divers zoom past disjointed wreckage with hoses looped out like small animal snares, far beyond the reach of the shoulder. I call this the "hoola-hoop effect."

Although manufacturers are reluctant to advise tampering with their products, and often disclaim liability for any use or alteration that does not conform to advertised applications or printed instructions, the innovative diver should realize that any piece of equipment is subject to modification in order to suit specific needs. Before you try to reinvent the wheel and make your own modifications, consult your local dive shop for advice.

In the case of regulators, you need only enough hose to enable you to turn your head freely without restriction; any more hose is a detriment. Because tanks can be worn high or low, because they can be carried as either singles or doubles, and because people have different body proportions, the standard hose length that comes with the regulator is not always optimal. The simple solution is to replace the hose with one of the proper length.

Sometimes, a hose might be long enough if only it swung around at a better angle. A swivel block can make this happen; it is screwed into a first-stage low-pressure port, and the hose is then screwed into the swivel. Not only does this permit the hose to rotate in a complete circle, but the block conveniently alters the angle at which the hose leaves the first stage.

The diver at left has an awkward hose arrangement that could be relieved by the use of swivels. Below is an example of the "hoola-hoop effect." Because of a diver's point of view, it is sometimes difficult to see how his own gear fits; an alert instructor can point out such deficiencies and suggest alternative configurations.

Left: An octopus regulator with one hose longer than the other. Middle: A mouthpiece plug clipped to a BC buckle. Right: Bill Baxter glued a magnifying lens to the inside of his mask so he can read his gauges.

If your regulator is rigged with an octopus, you should have a longer hose for the backup regulator you intend to pass to your buddy, so you don't have to stay so close together during an emergency. This regulator cannot be allowed to dangle; it should be snugged against your body, but it should not be tucked into a BC pocket or some other inaccessible place: it must be available at a moment's notice. The second stage can be hung around your neck on a detachable strap that loops over the mouthpiece, or a mouthpiece plug can be used and clipped to a utility D-ring. The last thing you want is to have a mouthpiece full of sand or mud when you or your buddy need a quick breath of air.

An alternative school of thought holds that a panicked diver in an out-of-air situation will snatch the regulator out of another diver's mouth rather than search for or ask for the backup, that the out-of-air diver will not be in any frame of mind to then exchange the main regulator for the backup, and that the assisting diver must thenceforth breathe off the backup. In this scenario, the recommendation is that the main regulator hose must be long enough for another diver to use.

For air sharing situations, a side exhaust regulator is preferred because it can be rotated when passed to another diver, allowing both to breathe while facing each other; this facilitates exchanging hand signals and reading gauges, and helps maintain eye contact (important during an emergency). An unfavorable circumstance is created by a bottom venting regulator because it cannot be flipped over; the hose must be bent back on itself for someone else to use it. This puts a strain on the hose and forces both people into a wrestling match that makes it difficult to communicate, swim, and ascend, and which exacerbates stress.

Furthermore, because a side exhaust regulator can be breathed upside down, you can invert your body in order to poke your head into holes and to look under hull plates; you don't have to worry about water pouring into the bottom vent and choking you on the next inhale. Only bottom venting regulators equipped with a diaphragm exhaust system will prevent the intrusion of water when the second stage is inverted.

(Too large a mouthpiece can cause fatigue, and is difficult to get into your mouth during a switch, particularly in an emergency.)

High pressure hoses come in varying lengths, as well. Too often I've watched people struggle to get a submersible pressure gauge to where they could read it: they tug on the hose and twist in the backpack in wild contortions. It's a simple matter to buy a longer hose in order to make the gauge more accessible for viewing.

Submersible pressure gauges come with a swivel seat so they can be turned toward your face for easy viewing, thus offering universal accomodation for all regulators despite the placement of the high pressure port. Occasionally, the gauge will unscrew from the end of the hose. Get in the habit of always spinning the pressure gauge clockwise, the direction that tightens rather than loosens on standard threads. Also, it's a good idea to clip off the pressure gauge with a strap or an elastic band so it does not drag along the bottom or catch in the wreck.

If you see tiny bubbles escaping from the rubber sheathing of a hose—any hose—replace the hose immediately. Despite the obvious safety hazard should the hose fail explosively, the cost of a lost dive is more than the cost of a new hose. Hose protectors will extend the life of a hose by spreading the area of stress over a longer length.

Some people find their gauges hard to read for another reason: far-sightedness. If you shadow-play the trombone while reading menus and labels on pill bottles, you should consider gluing a magnifying lens to the lower portion of your faceplate: your mask then becomes an underwater bifocal. A plastic dime-store magnifier won't do; get one with optical quality glass and a flat surface that is designed for the purpose. Dive shops carry them or can order them.

On the other hand, if you are near-sighted you might want to purchase a mask with a prescription-ground faceplate, or have corrective lenses glued to your existing faceplate. The problem with lenses in masks is that if you break the mask or lose it, you can't borrow a mask to make the dive. On the good side, a friend of mine once lost his glasses on a dive boat and had to drive home wearing his dive mask: he was a real hit at the toll booths.

A side exhaust regulator permits swimming inverted: no need to worry about choking on water pouring into the exhaust tee's bottom vent.

I have opted for wearing contacts. I don't have to worry about salt spray or fogging, breakage or loss, and I can use any mask I choose. I prefer extended wear lenses because of my tendancy to nap on the boat between dives, or to sleep on the way in. Insert your lenses before boarding and leave them in all day. Carry a storage case and cleansing solution in case you have to take them out, and a compact mirror if you want to put them back in. Most dive boats are clean but few are sterile, so take appropriate precautions. Remember, too, that the seas can get rough. One time, the boat was pounding so hard that I could not hold steady enough to get the lens to lift off my finger: I either jabbed myself in the eye or missed it altogether—fortunately, the dive was cancelled and we returned to the dock.

There is always the possibility that your contacts can get washed out if your mask floods. It goes without saying that you should keep your eyes shut until you clear your mask. If, however, a contact does get washed or bumped out of your eye during a dive, just take special care to keep the mask sealed to your face and chances are that the errant contact will survive the trip to the surface without getting lost. In case you do lose a lens, it's a good idea to pack an old pair of glasses in your dry clothes bag; they don't have to be your most recent prescription.

A mask can sometimes get bumped askew or knocked completely off, either from your buddy's arm or fin tip or from swinging your head into wreckage outside your field of view. I like to wear the mask strap under my hood as an extra precaution. That way, even if the strap breaks, the mask will not fall away and I can easily reseal it and clear it without having to grope for it blind. If your mask ever presses too tight against your face, exhale a tiny bit of air from your nose to equalize the mask, and the pressure will be relieved.

After many years of diving I have become conditioned with a Pavlovian reflex: I automatically salivate as I reach for my mask. Defogger can be bought in a bottle, of course, but I find it cheaper and more convenient to carry my own in its natural container; I have an endless supply and I always know where to find it. Although spitting is considered unacceptable behavior in polite society, it is as permissible in diving etiquette as it is in jogging. Be careful of post-nasal drip, however—mucus tends to clot and it doesn't defog as well. As a point of interest, due to different chemical make-ups, some people make better defogger than others.

It's important to keep a snorkel with you at all times. Most people carry it attached to the mask strap where it is readily accessible. This is okay if you use it a lot, but it also gets in the way. If it snags on wreckage in passing, it can dislodge your mask. If you carry it only as a backup, consider keeping it in your goody bag or tucked under your knife straps; it may not be as easy to get to, but it will be there in case of emergency, such as a long surface swim after you've exhausted the air in your tank. Then you can pull it out and slide it under your mask strap to hold it in place.

Dive knives are generally a cut above the average pocketknife. They are made of stainless steel to retard rusting, and come with a firm plastic or rubber

Nets are a constant hazard in wreck diving. Notice the spinal column in the sand under the fine net on the left, and the fish caught in the heavy-duty trawler net draped over the engine on the right.

grip with finger indents for good handling. The misconception about dive knives is that they are weapons, intended to fight off dangerous marine animals. Some people buy the biggest, baddest looking knife they can find because it looks macho. In reality, however, a dive knife is a tool.

The primary purpose of an underwater knife is to cut line: rope, netting, and monofilament fishing line. This is because the primary hazard peculiar to wreck-diving is entanglement. A good knife has other uses as well: for chopping up sea urchins to feed the fish (frowned upon by environmentalists), for stabbing flounder for dinner, for prying abalone off rocks, for signaling other divers (by tapping it on your tank), and for spreading peanut butter on bread when you're having lunch between dives. Although they're hard to come by these days, I like a knife with a solid steel shank that protrudes from the hilt so I can use it as a hammer; I use the blunt edge to knock encrustation off metal so I can see if it's brass.

Above all, buy a knife that doesn't cost a fortune because chances are that sooner or later you'll lose it: when it falls out of the sheath, when the straps tear due to wear or dry rot, or when you drop it in a klutzy maneuver because you're wearing thick neoprene mitts. Knives are expendable, so get one that is worth losing.

Wreck-divers carry at least two knives, a primary and a secondary, and they are carried in different places. Getting wrapped up in fishing line is apt to make you nervous and give you a case of the dropsies at a time when you can least afford it. Wear the primary knife on your leg, either calf or thigh, and make sure you're limber enough to reach it. Keep a secondary knife on your other leg, on your console, on your backpack strap, or on the harness of your buoyancy compensator. That way, if you get entangled in such a way that one knife is inaccessible, you can still reach your other knife. Ideally, one knife should be accessible to each hand, in case one arm gets hung up.

For a secondary knife I use an inexpensive kitchen knife like a Laser or Ginsu, the kind advertised as cutting a quarter-twenty bolt then slicing a tomato. They are so sharp they are dangerous. I snap off the point and grind it smooth with a file or a grinding wheel—otherwise the point will go through the bottom of the sheath and into the flesh; remember, this is a line-cutting tool, not a stabbing weapon. They don't come with sheaths, so you'll have to

I keep a small Ginsu knife strapped to my console; in the picture at left I am about to tighten a plastic cable tie around the sheath extension under the handle. At right, a large dive knife doubles as a prying tool.

use a dive knife sheath. And if you manage to keep it long enough the blade will rust; when that happens, trash it and buy another. Alternatively, there are some very good but inexpensive dive knives that also fill the bill: not only are they small, but they come with a snap-lock sheath instead of the traditional sheath with the rubber thong that eventually breaks or dry-rots.

There's a trade-off with knives: the better the quality of the stainless steel the worse it holds an edge. The lower the grade of the alloy, the sharper the blade and the quicker it rusts. Decide which is best for you.

Clanging tanks with dive knives is going out of vogue: not only does it dull the blade of the knife and chip the paint off the tank, but the metal blade doesn't make a very loud sound against the plastic mesh tank protectors now so widely in use. For sonic signaling, some dive shops sell a plastic ball attached to a bungee cord that slips around the tank; you simply pull the ball away from the tank and release it to make it clang. Because the ball is smooth it is easier on the paint, and it is small enough to fit through the weave of the mesh.

Another handy tool for cutting thin line and monofilament is a finger grip device designed to cut away parachute chord. It is quick and easy to use, comes with its own sheath with a Velcro flap, and can be affixed to just about any width strap or harness material. The plastic blade holder is shaped like a hook, with the blade on the inside of the hook; you wield it by pulling it toward you over the line.

Because shipwrecks are three dimensional, sometimes presenting high profiles for exploration at a variety of depths, buoyancy control is essential. This control is provided by a BC (buoyancy compensator), also called a BCD (buoyancy compensating device). Since wreck divers have a tendency to carry more equipment than the average open water diver, the most useful BC is the one with the largest capacity that will fit you properly. Buy the biggest BC that is comfortable to wear, and become proficient with the power inflator so you can adjust your buoyancy quickly without having to stop what you're doing in order to inflate it orally. This can save many minutes during the course of a dive, reduce your energy output, help you to maintain a normal breathing rate, and allow you to focus your attention on matters at hand without constant interruption.

If the BC comes with pockets you can use them for storage, if it has D-rings you can clip off accessories. If the BC does not have D-rings but has a waist or shoulder strap, you can usually add a D-ring by looping it through a metal slider. Do not trust valuable accessories to a plastic D-ring; they can break with a sharp, sudden tug, such as when you jump off a boat and the accessary momentarily gets hung up. Stainless steel D-rings are preferred.

Plastic sliders can also keep your lead weights in place on your belt. Loop the belt through one slider, then through the slot in the molded lead, and through another slider. Snug both sliders tight against the weight and it won't be able to move.

Tools should be wrapped with tape that is yellow, day-glow, or reflective, so they are easy to spot if you drop them in the silt. I keep an inner tube band on my hammer handle, chisel, and drift pin punch, as a lanyard; this enables me to let them go without losing them if I need a free hand to pick up another tool, redirect my light, or manipulate equipment. Another basic wreck diving tool is a crowbar, also called a pry bar or pinch bar. The length in common use is about two feet; it is usually kept out of the way by strapping it to the scuba cylinder.

I cover the cutting edge of my chisel with friction tape, to prevent it from slicing through the mesh bag. Once, as I prepared a new chisel on the boat by carefully folding several wraps of tape over the edge, my good friend (who shall remain nameless) asked me how I was going to get the tape off under water while wearing thick gloves. I said, "By hitting the blunt end with a hammer."

The single most important safety item essential for wreck diving is an alternate supply of air. I'm not talking about an octopus regulator, which accesses the air in your tank if your other second stage fails, but a fully redundant source of air with its own delivery system. It's a smaller version of the scuba cylinder called a "pony bottle."

Ask yourself these questions: can I reach the surface on the air in my lungs if my regulator shuts down after an exhale? Will I have enough air in my tank at the end of a dive to disentangle myself from a net? Can I hold my breath calmly for several minutes when I get stuck in wreckage and my regulator gets pulled from my mouth? If the answer to any of these questions is no, then you need a pony bottle.

The fact that you can switch at a moment's notice to a backup regulator that is hanging on a strap by your throat, usually enables a person to control his panic in a situation of stress. The extra air gives you time to compose your thoughts while you ascertain the problem, and helps you make good your escape to the surface while breathing calmly and freely and making a controlled ascent. The pony bottle is not intended to carry additional air that is factored into your total supply for the purpose of extending bottom time; it is emergency reserve air only.

The minimum size pony bottle recommended for wreck diving is fifteen cubic feet. Anything less offers only a false sense of security. Put a good

The clamping arrangement at left secures a pony bottle to a single tank. The diver on the right demonstrates how a pony bottle nestles snugly on the back of a double tank setup. It is better to test new equipment at a quarry or some other controlled environment, than to learn at sea that it does not fit or work properly.

regulator on the bottle so you know it will deliver air when you need it; and switch regulators regularly between your main tank and pony bottle so they get used evenly. And remember the cardinal rule: don't go diving without it.

Many people begin diving by renting gear from a dive shop. While this may seem like a waste of money when you could purchase outright and save rental fees, it has advantages. Renting enables you to try out different types of equipment from different manufacturers, and helps you decide what suits you best. In the long run it may save you money because you can make more informed purchases, instead of buying first, then discovering you really don't like the item or that it doesn't work for the intended application. On the down side, if you consistently rely on renting a wetsuit, you may find that your size is not always available.

Eventually, you need to have all your own equipment. Rental gear needs to be adjusted every time you rent, whereas your own gear fits you every time. Your own equipment can also be partially pre-assembled to save you time in dressing: your knife may live on your buoyancy compensator, your weight belt will have the precise amount of lead, your regulator hoses will be the proper length and come over the shoulder you find the most comfortable.

Not only is having your own gear more convenient, you quickly become acquainted with it and with its location on your person. Everything will always be in the same place, and dressing will become routine. You can reach over your head and trace a lost regulator hose because you know from constant use where it is supposed to be. You will be more relaxed in the water because instead of wearing strange gear rigged in unfamiliar ways, it all conforms to your body and is in easy reach. Soon your dive gear will become part of you; donning it will become second nature, and you will swim through the water with the agility—if not the speed—of a fish.

Take good care of your dive gear and it will take care of you. Rinse it thoroughly with fresh water as soon as possible after an ocean dive. If you let it go till the next day, it's a good idea to let it soak for a few hours in order to dissolve encrusted salts and minerals. Just fill up the bathtub with tap water and plunk everything in. Get a watertight cap for your regulator's first stage, but remember that when the second stage is not pressurized water can get sucked up the hose if you press the purge. Pour water from the spigot into the hose of your buoyancy compensator, drain it completely, and repeat the process; that will get the salt out of the inside. Afterwards, lay everything out to dry—but out of the direct rays of the sun.

Whenever possible, soak your wetsuit in a solution of water and wetsuit shampoo: thirty gallons of water to half an ounce of shampoo—there shouldn't be any soap bubbles. This not only breaks down salt deposits that damage the neoprene, but it kills odors as well.

My grandmother was a wise woman who used to say, "An ounce of prevention is worth a pound of cure." Such rural sentiments are as true in the world today as they were on the farm a generation ago. Every moment spent working on your gear at home will pay for itself on the bottom where it counts. Unfortunately, many things don't malfunction till you use them. (When was the last time your car broke down in the driveway?) Nevertheless, look over your gear for breakage, stress cracks, frays, rust, and signs of wear. Perform maintenance and make repairs as needed: don't wait till the next dive when it may be too late, or after you've forgotten. Do it right away. If the work required is beyond your capability, drop into your local dive shop and have it taken care of at once, or ship the item to the manufacturer for repair. Remember, when your car breaks down you only become a pedestrian; but diving is like flying: a system failure results in a crash.

My motto is: a place for everything and everything in its place. I used to stow my gear on shelves in the basement, until I found a cocoon in the first stage of my regulator; I gagged at the thought of having a moth or butterfly purged down my throat at the first breath in the water. Store your gear in a sturdy zippered bag or lidded box. Not only will this ensure that your gear is packed for the next dive, it will help prevent you from leaving essential items behind, from losing them at the dive site, or getting them mixed up with other people's equipment. Because it is hard to distinguish one person's gear from another's, you might want to brand your equipment with an indelible marker.

Remembering that for the want of a nail an entire kingdom was lost, a dive may have to be aborted by so simple a plight as a broken fin strap—which will never break until you stretch it over your heel. Put together a spare parts kit that contains such perishable items as a mask strap, regulator mouthpiece, and o-rings; throw in a few simple tools like screwdrivers and pliers; and include such temporary mending materials as heavy-duty rubber bands, plastic tie wraps, and the diver's cure-all and panacea of jury-rigged repairs, duct tape. Depending upon how much stuff you carry, you can keep it all in a press-seal plastic bag, a zippered pouch, a snap-lid food container, or an all-purpose watertight case with a gasket around the lid. Some people call this a Murphy

One half of my double-sided parts kit.

box, named after Murphy's Law: "If anything can go wrong, it will." I use a sectioned tackle box not only so I can separate my goods, but so everything is visible and easily accessible without having to root through a "junk box" in which the item wanted is always on the bottom.

As time goes on and your expertise grows and you formulate new interests, you may decide to upgrade or update your equipment. I like to think that dive gear is only a temporary possession. You will find a ready market for items for which you no longer have a use: there is a steady stream of divers following you on the equipment treadmill, and they will look at your gear as an upgrade on what they already have. If you're in a rush to sell, you might want to publish an advertisement; but if you can afford to wait, word of mouth will eventually reach those who are in need of what you wish to recycle. The income will help defray the cost of new purchases.

In later chapters I will discuss other specialty items such as lights, line reels, compasses, and cameras: to use under certain conditions on wrecks and to enable you to achieve definite goals.

"Darn! I knew I should have bought the extra large bug bag."
(From *Twenty Thousand Leagues Under the Sea*, circa 1872.)

Physical Protection and Thermal Considerations

There is probably no feeling as free as that of diving into the ocean wearing nothing more than a bathing suit or a pair of swimming trunks. Since your arms and legs are unhampered, you have unrestricted motion for swimming and kicking, and, if the day is hot and the sea tropical, the water will wash over your body with a soothing and refreshing coolness. With the addition of mask, fins, and snorkel, you can easily glide over wrecks sunk on shallow reefs and observe from above the chaotic structure and attendant marine life. Intrigued by the panorama unfolding below, you can forget your worldly troubles; the time will pass pleasantly.

Then, you return to the realm of open air and find that the skin on your back has been burnt to a crisp. Even if it's only a bright cherry red, you're in for several days of unrelenting agony.

This scenario is quite common with northern folks who visit the Caribbean after a winter of hard indoor activity. It's easy to forget that bleached white skin is vulnerable to only a few minutes exposure to the rays of the sun. More insidious is the overcast sky, when clouds prevent the passage of low-energy infrared radiation (which causes heat) but present no barrier to high-energy ultraviolet radiation (which tans the skin). Add the cooling effect of water and we have a recipe for disaster.

The best prevention for sunburn is nicely tanned skin. Nature's way of curing the epidermis is through gradual exposure. But in today's cavelike society, people live and work largely within the confines of air-conditioned buildings, and exposure to the sun is either unsystematic or patchy. Even avid boat divers tend to get what I call "diver's tan": coloration only on the face and hands because the rest of the body is sheathed in neoprene.

I've been known to wear thin pants and a long-sleeved shirt while exploring shallow wrecks in tropical conditions. Not only does the material offer protection from the sun, it helps prevent minor scrapes from contact with the wreck and from sharp marine organisms such as coral and sea urchins.

Nowadays "skins" are all the rage. They are sleek, colorful, quick drying, form fitting yet stretchable, appropriately yuppie, and surprisingly useful: in short, a fad that is here to stay. They have become an accepted form of dive-boat dress, the tuxedo of the underwater world, the pinnacle of sartorial perfection. Their main appeal is fashion: they come in a vast array of dazzling color schemes, and they accentuate the wearer's physical attributes— good for some, bad for others.

Skins are one-piece suits made of sturdy abrasion-resistant materials that offer some defense against encrustations, safeguards against chafing from tank straps and weight belts, can ward off the stings of jellyfish and fire coral, and makes life easier when annoying insects are buzzing around the boat. The suits are often called "skins" because they are so comfortable you are hardly aware of having them on. And since they cover you from neck to ankle, you need apply sun block only to your face, hands, and feet.

Polypropylene skins are warmer than Lycra-nylon, but don't come in all the jazzy colors. For better warmth and protection, a hood, gloves, and socks can be added to the ensemble. They are also available in a thicker, more durable style that offers thermal protection: for those days when there is a chill in the air, a breeze across the sea, or cold water currents driving through the reefs. For colder conditions, a thick skin suit can be worn under a wetsuit specially designed for the "layered" effect. The advantage of such a system is that you don't always have to wear that bulky, quarter-inch wetsuit because it's the only one you own. The combination of a thick skin and a thinner wetsuit enables you to suit up for the super cold or dress down for milder conditions.

What I like best about skins is how they help me slip into my wetsuit. In the old days of nylon-1 I had to apply liberal amounts of corn starch or baby powder to the inside of the suit, with varying degrees of success. Cornstarch and baby powder work well when they are dry, but if your wetsuit is damp you're likely to work the powder into a paste. Besides that, when you sprinkle the powder down the legs and sleeves, the stuff blows all over the deck and other people's gear. Wafted in the wind, I've seen dive boats wind up looking like bean fields after an aerial crop duster dropped its load. Powder doesn't work with the nylon-2 "plush" suits on the market today.

Liquid dishwashing soap is a suitable compromise, and I've used it for years. Don't buy a no-frills brand; get one that is mild on the skin because you're going to be soaking in it for quite a while. Harsh soaps can cause a rash or dry out the skin. I once had flakey arms and legs for two weeks after a dive trip; I itched the whole time. And never use soap full strength; dilute it with water at least fifty-fifty, more if you have sensitive skin. If you carry it in its original squeeze bottle, make sure the top is screwed down securely or you'll have a gear bag full of suds. Baby shampoo has an advantage in that it won't sting if you accidentally get it in your eyes.

The main difficulty with using soap is finding the proper amount to squirt into the suit: too little and you don't get the glide, too much and you foam around the cuffs and ankles like someone with a weird case of rabies. You also need to rinse off after removing your suit; you don't want that soap drying up on your skin because it will induce chapping. There's nothing quite like jumping naked into the water after an unpleasantly cold striptease act on an open boat.

If you plan to dive mostly in colder environments the layered wetsuit style may not do you much good; it will simply increase your dressing time because you need both layers all the time. In that case you'll want to get a heavy-duty wetsuit, and the warmest you can buy.

Some people are lucky: they grow into generic body shapes that require no measurements other than small, medium, or large. In that case you can buy a standard "rack" suit as if you were ordering a pizza: right off the shelf without any alterations. What toppings did you want?

Most people, however, are not so shapely nor as easily accommodated; they need alterations to a rack suit or a custom made wetsuit to account for personal variations in height, girth, and limb length. This is not so they look good on the wreck, but so they feel comfortable throughout the dive. In order for a wetsuit to do its job properly, it needs to fit snug but not too tight. The importance of fit can be clearly understood by examining how a wetsuit works.

A wetsuit keeps you warm by trapping water next to your body which is then heated by the body. The trapped water and the thickness of the material insulate you from external ambient temperature. If the wetsuit is too loose, water circulates freely and is continuously flushed out with unheated water from outside. As long as cold water is washing across your body, you may as well be wearing your birthday suit.

By extrapolation, it would seem that the tighter the fit the better the suit. This is true only to a certain extent, then other factors enter into the equation that offset the degree of warmth; anyone who has worn a shirt or blouse with a choking collar can understand what I mean. A wetsuit that is too tight is overly restrictive and potentially dangerous.

What! No tie! The diver below is wearing black pants, a white shirt, and gloves, as physical protection in 80° water. The divers at left are displaying the partial look: wearing half a wetsuit, either top or bottom.

A good rule of thumb is: if you have to fight your way into a wetsuit, it's too tight. Too much struggling saps your energy reserves, exhausting you before you get into the water. On a hot day you can get seriously overheated, dripping beads of sweat that deplete your store of water, which in turn diminishes your body's ability to metabolize food. If the material bunches up at pressure points (the neck, armpits, and crotch) it will restrict the flow of blood, causing headaches and "pins and needles" in the arms and legs, and put you at risk of incurring decompression-related problems. A too-tight wetsuit forces you to stretch rubber whenever you flex your arms and legs, bend over or crouch, or simply walk around. Who needs to do isometrics when you're saving your strength for the dive? Furthermore, stretching the neoprene not only results in a thinner material layer, it causes the suit to wear out faster.

There are ways to maximize the warmth provided by a stock wetsuit. Because the largest loss of heat is conducted through the head, a hood is essential in cold water. Tuck the extended collar under the jacket to protect the neck, and to prevent those icy fingers of water from jetting down your back. Or, if you buy a jacket that is a little bit loose you can wear a hooded "chicken" vest underneath it: without sleeves it is nonrestricting, while the extra layer of neoprene adds to the comfort level, and the attached hood prohibits water flowing down the wetsuit collar from contacting your skin.

Gloves and booties will not only keep the hands and feet warm, but the overlapping material will significantly retard the ingress of water. Roll up the pants cuffs, smooth the tops of the booties up your legs, then roll the cuffs down over the booties; this way, as you move forward through the water the bootie tops won't flare open and suck in water.

Common among cold water divers is the trouser style known as "farmer johns," in which the pants extend above the hips and all the way over the shoulders. The addition of a jacket affords double thickness protection for the chest and vital organs, and retards the circulation of water around the waist. A spine pad fills in the gap for people with a sunken backbone; this prohibits cold water from channeling in from the top or bottom and causing those spine-tingling chills.

Wetsuits are rugged and easy to maintain. A thorough fresh water rinse after each day's use is the minimum care required, but as mentioned in the previous chapter, a long soak in wetsuit shampoo will dissolve encrusted salts and minerals, and will wash away accumulated perspiration and other body odors. Hang the wetsuit to dry, and leave it hanging until you pack it for the next dive. Wetsuits folded for extended periods of time tend to get permanent creases which damage the neoprene cell structure and create thin spots, both of which adversely affect the material's ability to retain heat.

Wrecks and coral can wreak havoc on wetsuits, tearing the material and digging deep gouges which, you must appreciate, could have been damage done to your skin. Do not make repairs until the wetsuit has dried completely; and make sure the inner surfaces of the damaged areas are dry before applying neoprene cement according to instructions on the container.

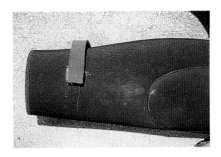

A knife retainer glued to the leg of a wetsuit.

If you see loose threads, nip them off with a pair of scissors and place a dot of glue on the spot in order to prevent the thread from unraveling. If the seams split at the neck or cuffs, and need to be reglued, slice off the exposed edges with a razor blade so you have two clean mating surfaces. Plastic zippers need no extra care, but a little beeswax or parafin wax can improve the glide of metal zippers.

Wetsuits can be modified by gluing on pockets, tool pouches, and knife retainers. I always place a thin strip of neoprene over my sheath and bond the ends to the leg; it prevents the sheath from moving out of position as the thickness of the material compresses due to depth.

Compression causes a few other problems that need to be addressed, hood squeeze being the most serious. As you descend, the water presses the material of the hood hard against your head. If the hood has a very good fit, it can trap air that will compress against the ear drums. If you feel excessive pressure, pull the hood away from your ears and let the water flow in and equalize. If hood squeeze is a chronic problem, poke holes in the sides of the hood so water can enter the ear cavity freely.

Another manifestation of compression due to depth is the loss of buoyancy as the neoprene cells are crushed into a smaller volume. This means that you need to add air to your BC on the way down, and to vent air upon ascent lest you rise too fast. Worse than this minor inconvenience and constant attention to buoyancy control is the fact that as the material gets thinner, its insulating quality is correspondingly decreased. Not only does the water get colder below the thermocline, but your wetsuit retains less warmth: a double loss of thermal protection when you can least afford it.

While we're discussing wetsuit drawbacks, let's not overlook the downside of being encased in neoprene too long before the dive begins. The most efficient dressing sequence is to rig your regulators, arrange your accessories for quick deployment, don your wetsuit, get into your scuba unit (tank and BC), grab your prearranged gear, and enter the water. In practice, dressing seldom works with such assembly-line precision.

If you're diving from the shore, you may have a long walk ahead of you. If it's a boat dive, and you try to anticipate arrival time, you may get fully dressed, then discover that the captain can't find the wreck or that the grapnel

A wetsuit is desirable in shallow tropical waters, not only to prevent nasty scrapes with the wrecks (what we call "wreck rash"), but because most people will get cold after extremely long exposures.

keeps missing or slipping out. I once spent an entire day suited up only to return to the dock severely dehydrated and without ever making a dive; it happens. In the hot sun, every minute sitting in a sweaty wetsuit can be downright painful.

This might make you think it's better to wait until the grapnel is secure before putting on your wetsuit; but the delay puts you into the water later and cuts into your surface interval. Sometimes, postponement is unavoidable, such as the days when so much spray is cascading over the deck that if you kept on your street clothes while preparing your gear, you'd get soaked, and have to spend the return trip cold and miserable. In that case, you're better off donning your wetsuit first before wading into the elements to prepare your gear. Make a judgment based upon conditions and the way you feel.

These timing problems are exacerbated considerably if you compensate for extremely cold bottom temperatures by going to a three-eighths-inch wetsuit. Not only are you likely to swelter miserably on the surface, but your mobility will be severely curtailed. It takes a strong person to bend that much rubber.

If you're really committed to diving in near-arctic conditions, there's only one way to go, and that's dry. Drysuits come in a variety of types, materials, styles, and zipper configurations. Basically, a drysuit is any kind of diving dress that prevents water from contacting the body; warmth is provided by the combination of the insulating characteristics of the material of which the suit is made, and the amount and quality of clothing worn underneath. By this definition, the hardhat diver's canvas suit (with breast plate, helmet, and lead boots) constitutes a drysuit. So do the World War Two varieties in which the diver entered the suit through the stomach, then wrung the stomach material together and tied it tight like you would wrap a twist-tie on a cellophane wrapper. But in wreck-diving parlance, the generic drysuit is one that you wear like a pair of footed, one-piece kiddy pajamas (without the rear flap).

Drysuits are made out of neoprene, nylon, or vulcanized rubber. The features common to them all are a waterproof zipper to permit entry, and seals at the neck and wrists. Thus the only parts of your body exposed to the water are the head, face, and hands.

A few drysuits come with attached hoods, some of which seal around the face and keep the head dry. Some hoods are made of neoprene, but if they are made of latex, which has little to no insulating capacity, a liner is worn underneath for warmth. If you procure a drysuit without a hood, or don't like the liner for the latex hood, get a wetsuit hood to protect your head. For extreme conditions or dives of long duration, or when wearing a drysuit whose attached hood is made of latex or crushed neoprene, I wear two hoods: the one that comes with the suit, and a cold-water hood that seals around the face and over the chin, and that has connective material that covers the upper lip. The extended flaps on cold-water hoods make mask adjustments difficult because the overlapping material must be pulled out while the mask is being sealed against the face and forehead; but when the flaps are rolled over the edge of the mask, the skin is completely covered and therefore protected from the cold.

However, a note of caution must be appended to the use of cold-water hoods. If the upper lip flap is too wide, or if it is positioned improperly due to a poor match between your facial configuration and the hood design, the rubber flap may have a tendency to pull down. Then, if you remove your regulator, the flap will slip across your mouth and prevent re-insertion of your

An early prototype drysuit.
(From *Twenty Thousand Leagues Under the Sea*, circa 1872.)

regulator. It's easy to pull the flap out of the way *if you ascertain the problem* and *if you don't panic*. I know of one fatality that was attributed to the lip flap blocking the mouth. To avoid this predicament, I cut my lip flap back with a pair of scissors until my lips protrude freely through the oral opening.

If I wear two hoods, I wear the mask strap over the inner hood and inside the outer hood. As mentioned in the previous chapter, this lessens the chance of losing the mask.

Drysuit hoods have a tendency to trap air that is "burped" up the neck seal or exhaled through the nose and out the top of the mask. The hood balloons, making it wet and uncomfortable, so you have to continually squeeze out the air by pressing the hood against your scalp. Most people alleviate this annoying problem by cutting two or three small escape holes in the top of the hood. Also available are hoods with a check valve: two flaps of neoprene with offset holes that let out air without letting in water. A check valve can be installed on an existing hood.

Wetsuit gloves are sometimes worn with a drysuit, but usually, if it's cold enough to require a drysuit, three-finger mitts are preferred. There's a slight loss of dexterity, but there's not much need to have a fully independent pinky underwater. Some mitts are designed specifically for drysuits: they come with extended cuffs that fit snugly over the wrist seals of the sleeves, thus preventing or at least limiting water intrusion. If your mitts tend to leak at depth, it's better to flood them at the surface before dropping through the thermocline.

A perfect seal between wrist and cuff keeps your hands dry, but presents a slight problem of its own. If you hold your hand over your head, air is likely to "burp" from the sleeve into the mitt. The inflated mitt expands, and you're left wiggling your fingers inside a mitt that has suddenly become too large and whose fingertips bulge way beyond the ends of your fingers. Now you've got all the dexterity of a boxing glove. The solution to this problem is to hold the affected member down by your side, squeeze the mitt with your other hand, force the trapped air out through the cuff, and work the finger sheaths back into position.

Drysuits with integrated cuff rings and snap-on dry mitts suffer the same

Jon Hulburt illustrates two ways of adding warmth to the wet parts not protected by a drysuit: two hoods provide extra insulation for the head, and a blast of hot air from the dive boat's exhaust pipe heats the inside of his mitts. Some people pour hot water from a thermos into their mitts.

problem of unwanted inflation; the above solution applies. Some manu-facturers recommend placing a soda straw in the wrist seal to allow air to flow freely between the sleeve and the mitt.

There are several ways of extending the life span of neoprene mitts and gloves, and to compensate for uneven wear. I put a *thin* coat of AquaSeal on the palmside fingertips of my right mitt because that is my "grabbing" hand. (Since I hold my camera or light in my left hand, my left mitt gets less wear and tear from the wreck and from abrasive encrustation.) Too thick a coat makes the neoprene stiff and unbendable. I make sure to buy interchangeable mitts—those without left-handed or right-handed thumb positions. That way, I can switch them from hand to hand so they wear out at the same time, instead of building up a collection of perfectly good left-hand mitts. Nor am I above turning my mitts inside out when the outside material frays or shows signs of wear. By employing all these methods I can get over four times the use out of a single pair of mitts. When they really get ragged, I relegate them to summer use.

You can also extend the life span of a drysuit by taking some simple precautions, by practicing good maintenance, and by making a few modi-fications. Until someone invents a flexible kevlar drysuit with an affordable price tag, you're going to have to learn to treat your drysuit with care: care that begins at home, is carried out on the boat, and is continued underwater.

When not in use, keep your drysuit hanging so it does not take on creases. Don't use a wire hanger because the weight of the material will indent itself upon the slender strand of wire. Drape the suit backward at the waist over a length of PVC pipe hung by ropes from the rafters or some other suitable point of ceiling attachment. A cool, dry area like a basement or garage is preferable to outside in the sun. A heavy-duty coat hanger pushed up through the neck is not recommended because the neck seal might get torn in the process of insertion or removal.

Rinse the outside thoroughly after a salt water dive; you can lay it out on the driveway and wash it down with a hose. Use hand soap and a bristle brush to scrub out stains, but no detergents: the emulsifying chemicals are too strong for drysuit materials and can cause premature deterioration. The valves need special attention in order to ensure smooth operation and to prevent them from jamming open: take care to squirt water all around the buttons and up into the inflator nozzle. Do not depress the exhaust port or water will dribble inside. Also, make sure not to get water up the sleeves or down the neck, or you might have a surprise waiting for you the next time you go diving; evaporation in an enclosed watertight envelope is a very slow process.

Spray silicone on the valves to keep them lubricated, as well as into the power inflator connector on the hose and around the disconnect sleeve in the pulled-back position. However, most manufacturers recommend that silicone *not* be used to increase the glide of waterproof zippers; the hydrocarbon compound can weaken the material to which the zipper is bonded. Instead, wash the zipper occasionally with warm soapy water. If the metal teeth show green spots of verdigris, scrub the zipper with an old toothbrush. After the zipper dries, rub beeswax over it—not too much or you'll gum up the works; a little dab'll do ya.

Latex seals need to be rinsed inside as well as outside. Oils from the skin and salt from sweat cause the material to deteriorate.

The places where a drysuit gets the most wear are the feet and the knees. Shore diving is exceptionally tough on the soles, especially if the beach consists of sharp granular sand or jagged rocks. If you wear your drysuit from the dressing station to the shore, instead of donning it at the water's edge, you might want to pull on a pair of galoshes to protect the soles and sides of the booties from puncture wounds.

Even boat decks can be destructive to footwear because of cleats, screw heads, and splinters. More of a problem is scraping your booties across tools, tank valves, and rough-edged dive gear normally found littering a crowded dive boat. Walk lightly on the deck.

Despite hard usage on the soles, the spot on a bootie most likely to wear out is the top, where it scrapes across the foot pocket of the fin every time you pull it on. Despite this obvious and longtime weak point, some drysuit manufacturers persist in placing a seam on the front of the bootie precisely where it does the most damage: the threads soon fray and pinholes eventually develop. To guard against this problem, I always put a thick bead of AquaSeal on the front of the bootie wherever it contacts the fin pocket; it may not look nice, but it does the job.

You'll find that the foot pocket of a fin that fits well over a wetsuit bootie is usually too small for use on a drysuit, especially if you wear thick socks and insulated liners. Conversely, fins that fit a stuffed drysuit leave your foot flopping around when wearing wetsuit booties. If you dive both wet and dry you'll need two pairs of fins. In a moment of economy, I discovered that the hard plastic bootie tree, intended to maintain the shape of the foot pocket when the fin is stuffed into a gear bag, can be used as a spacer. I leave it in place when diving wet, and remove it when diving dry.

Left: A bead of AquaSeal up the seam of the legs will retard chafing. Middle: Tape is being used to make an emergency repair on a torn wrist seal. Right: Dave Poponi looking his best, while keeping out of the wind on a raw day in Nova Scotia.

Drysuits can take quite a bit of abrasion and rough handling. The dirt on Lynn DelCorio's suit (at right) bears witness to his bottom crawling exercises.

Although most drysuits come with glued-on knee pads, they are not always adequate for the rugged shipwreck environment. Wreck divers have a habit of resting on the bottom in the bent-knee position, pressing hard against metal protrusions and sharp marine encrustations that can puncture the material if enough force is applied. Even if the drysuit does not develop leaks at the knees, the pads can get pretty tattered and are costly to replace. I've invented my own, add-on knee pads (appropriately called "garypads") which are similar to those worn by carpet layers and cement finishers.

Garypads are cut from automobile inner tubes in sections about a foot long. Pick a size that's slips easily over the legs and stays in place without being too tight: somewhere between E-15 and L-78. Cut an oval out of the back, leaving a two-inch band at the top and bottom. The inherent curve fits over the knee joint nicely, and the rubber is flexible enough so you can bend your legs without feeling as if you're doing army calisthenics. Garypads have ·an amazing resistance to abrasion; even sharp pointed barnacles have trouble slicing through. They last virtually forever, and replacements are usually free: old inner tubes that no longer hold air for inflating tires are still good enough for garypads.

Lest this discourse on foot soles and garypads lead you to believe that drysuits are as delicate as a chiffon wedding gown, and must be handled with dainty concern, banish the thought. They are tough, can take a lot of punishment, but like any other piece of equipment they require care and maintenance to ensure longlasting use. Pinhole leaks are unavoidable, and their precise location can be difficult to detect without total submersion and minute inspection. An annual checkup at the factory or a reputable repair facility for leak testing and a plug job will enable you to sing that familiar commercial ditty, "How dry I am."

You can also perform your own checkups in the field. Those of us who dive dry all the time have developed a habit of noticing that telltale trickle of bubbles rising to the surface from our buddies' drysuits. We automatically make a mental note of the spot, quite often a seam, and report the information after the dive. "You watch my back and I'll watch yours," is the watchword of the wise.

You can usually check your own arms, feet, and the front of your legs. If you've got a known leak in an unobservable spot, show your buddy where you're getting wet and have him examine the area on the next dive. He can make a mental note of the where the bubbles emerge.

There are other ways of locating leaks, most notably by overinflating the suit and smearing it with soap or shampoo. Bubbles will form over the tiniest pinhole. You can do this with your person inside the suit while someone else works you into a lather, or by plugging the neck and wrists with appropriately sized food cans or plastic jugs, so you can apply the lather yourself. Whatever floats your boat.

You can also do home repairs. Because I travel quite a bit I've put together a drysuit repair kit consisting of AquaSeal, cotol (an AquaSeal accelerator), neoprene cement, an inner tube patch kit, and a hair dryer. All bonding surfaces must be completely dry in order for a patch to hold or glue to set; the hair dryer comes in handy when you have to make overnight repairs for the next day's dive. Turn the hair dryer on a low setting, prop it in front of the affected spot, and let it blast away—but not so close that it melts the material. Once sufficiently dry, pull or pinch the hole apart in order to expose the inner surface, and blow it dry. Then invert the suit and do the same to the other side.

For quick emergency repairs, squirt a blob of AquaSeal over the pinhole and let it dry; use accelerator if you are making a repair during surface interval and want to use the suit on the next dive. For extra insurance, repeat the process on the inside of the hole. The hair dryer can also be used to hasten the drying process.

For a more painstaking repair job on a neoprene drysuit, pull or pinch the hole apart, apply neoprene cement to the inner surface, release the material so the hole draws together, and smear a dollop of neoprene around the erstwhile opening. Let dry, or rush it with the hair dryer. Then follow the AquaSeal procedure from the previous paragraph.

Vulcanized rubber, nylon fabric, and trilaminate drysuits can be fixed almost instantaneously with a bicycle tube patch kit. Scuff up the surface around the pinhole with the metal scraper provided with the kit, apply glue to the suit material and to the patch, then press the patch in place and hold it until the glue sets. A bead of AquaSeal along the edges of the patch will prevent the corners from peeling. Except for vulcanized rubber suits, which are vulcanized only on the outside, repairs can be done on the inside if you wish to avoid the unsightly patchwork quilt effect of a well used drysuit. Just keep in mind that service is more important than beauty.

As a preventative measure, I apply a bead of AquaSeal to the longitudinal cuff seams of neoprene drysuits because they eventually tear apart from the constant stretching. When they tear back far enough water is bound to force its way up the sleeve. At the end of the cuff seal I extend the AquaSeal bead half an inch in either direction—*but never run AquaSeal all the way around the wrist* because it does not stretch. Sooner or later, all cuffs need to be replaced.

When you first get a drysuit you may find that the neck seal is too tight. They usually come that way because it's easy to make alterations for a larger neck. If the neck size is at least in the ballpark, try stretching it over a two-liter soda bottle for a day or so, and it might do the trick. Otherwise, snip away a quarter inch of the end of the seal and try it on again. If it's still too tight, or if you can't get it over your head even by lubricating it with silicone spray, soap, or talcum powder, keep snipping off a quarter inch at a time until you can wear it without choking. It's okay if it's a little tight at first; you won't have to put your head through it too many times before you stretch it into the comfort zone. Follow the manufacturer's recommendations if there are any.

The frustrating part about neoprene neck seals is that not long after they get comfortable, they begin to get loose. Then you have to take them in. You do this by cutting a deep "V" along the seam, about half an inch across the top of the "V," and cementing the two sides together. Push the sides in slowly from the bottom of the "V" and make sure they come together evenly at the top. Stitch it up from the outside without pushing the stitches all the way through the material, then run a bead of cement on top of the stitches. It takes a little finesse, and the job may not look professional when you're done, but it works. There's a limit to the amount of tucks you can take in the neck seal before there's not enough neoprene to give you the necessary stretch; then it's time for a new neck seal, and the process begins anew.

Nothing makes up for a properly fitting neck seal. Adding rubber bands or cinching straps around your neck is to be avoided at all costs. Such expedients could restrict breathing, cause choking, and might possibly lead to panic and even unconsciousness.

By now you may be asking yourself if a drysuit is worth all the time and effort—to say nothing of the expense. As a dedicated user for more than twenty years my opinion is undoubtedly biased. But when the difference is between shivering uncontrollably (when I'm barely able to keep my mind on the wreck and remember that I'm supposed to be having fun) and enjoying the dive in the luxury of relative warmth, I think any reasonable person would opt for the costly comfort of diving dry.

The first time you jump into a frigid sea and are not forced to gasp as icy fingers of water invade your wetsuit and course along your spine like a re-enactment of *The Tingler*, you will agree. No more stripping off a dripping wetsuit and suffering the evaporative chill of the slightest breeze. No more putting that cold wetsuit back on for the second dive. Instead you can relax on the boat in dry luxury, and unless the day is exceptionally hot you won't have to remove the drysuit at all, unless you want to pull your head out of the constricting neck seal or attend to unavoidable bodily functions.

Neoprene drysuits have some inherent warmth, while drysuits made of other materials rely on clothing worn underneath to provide all the warmth. This enables you to layer your clothes for anticipated conditions. For summer use, some people wear only shorts and a T-shirt, but to avoid that clammy feel I prefer to have my pores covered with something that will absorb moisture. Lightweight longjohns or a one-piece union suit will get you by. Feel cold? Pull on a sweater or sweatshirt. Drawback: not enough padding on the chest leaves me with a painful imprint of the plastic spacers on the exhaust valve. If your valve does not already have one, glue a neoprene pad on the inside, but be sure not to seal off the valve.

For super low temperatures I wear expedition weight polypropylene longjohns, an old woolen sweater with the arms cut off, a heavy duty one-piece Thinsulate undergarment, and a pair of thick socks with Thinsulate liners. I cut the arms off the sweater because I need the protection for my chest and vital organs, not my arms; too many sleeves bunch up and restrict arm movement. On cold winter days I will arrive at the boat already wearing my longjohns, and will go home with them still on; that obviates having to strip naked in subfreezing conditions on an unheated dive boat. Drawback: too much legging material makes it difficult to negotiate boat ladders. I sometimes climb up on my knees.

Between dives it is common to see people walking around in their underwear. Longjohns are soft, warm, and accepted boat dress: the most important part of a diver's wardrobe and what the natty diver wears.

Some people wear a thick pile undergarment called a "woolie." These are almost as warm as Thinsulate, and provide nearly the same amount of bulk (the air space between your body and the water has an insulating quality of its own), but loses much of its heat-retaining capacity if your drysuit leaks and it gets wet. It also becomes a sodden mass and takes forever to dry, and produces more lint than a cotton mill. There have been complaints of lint jamming open the exhaust valve, causing it to leak.

In tropical conditions the drysuit is more trouble than it's worth. You'll

Bill Schmoldt started out with a BC over his drysuit, but stopped wearing it because of difficulty in reaching the inflator valve. Notice, too, how nicely the vulcanized rubber suit conforms to his body. A snug fit reduces dynamic drag.

sweat to death on the deck, won't be able to cool off in the water, will have an unbearably hot swim back to the boat, and the suit will get chopped to pieces by the coral. Go wet, young man; go wet.

In temperatures from moderate on down, I opt for the drysuit. I wear it so often that it has become part of my diving system. The power inflator puts buoyancy control at the touch of a button, and, even though for liability reasons manufacturers warn against using a drysuit as a buoyancy compensator, most people do. In fact, all BC's except those that are back mounted are potentially dangerous because they cover the control valves. If the drysuit inflator valve jams open and you can't get to the exhaust button or reach the quick disconnect, you're in for a rocket ride to the surface.

Against manufacturers' recommendations, I also use the suit to offset the weight of small objects I pick up during my dives. If the drysuit has these capabilities, why not take advantage of them? Would you buy a passenger car and not use it to carry packages home from the store? Of course, if you plan to move a heavy steel safe you would want to rent a truck—or get a liftbag.

The advantages of being dry are more than just not having to towel off after a dive. In my mind, wet equates to cold and dry to warm. I use much less air when I'm not shivering, I can concentrate better on the task at hand, my time in the water is spent more efficiently, and I can enjoy myself and take pleasure from an otherwise hostile environment. You must take these factors into consideration when weighing the cost effectiveness of a drysuit against that of a wetsuit. There's more at stake than the initial layout of cash.

Now that I've danced around the peripheral issues, which is the better drysuit to buy? It depends upon how you define "better." Is a Rolex a better watch than a Timex? Is a Caddy a better car than a Chevy? The answers to these questions may seem obvious until we get more specific. Can a Rolex tell time better than a Timex, or can a Caddy get you anywhere faster than a Chevy? The common denominator in drysuits is dryness. Beyond that, there exist a multitude of comparative elements such as cost, comfort, warmth, fit, reliability, durability, and ease of proper maintenance. To a certain extent, the better drysuit for you may not be the one that is the most prestigious by name, but the one you can get a deal on. Economics count. However, keep in mind that when you buy too cheap up front, you might wind up paying increased repair costs farther down the line. The same is true if you buy used: make sure the investment won't cost you more in repairs and maintenance than the item is worth.

Let us return to the types of drysuits and look at the different features, so you can decide which you're willing to pay for and which you can dive without. Some peculiarities that fall into the personal preference category are self-donning capability and the placement of zippers and valves. I don't like an exhaust valve on the upper shoulder, even if it vents automatically, because it can be reached only with the right hand. Too often I find that hand encumbered with gear or with hanging onto the anchor line, making it nearly impossible to operate the valve manually without awkward maneuvering. If the location of a valve is not to your liking, it can be moved and the hole can be plugged. If the position of the inflator nozzle puts a strain on the hose, the assembly can be rotated.

If you need help with your zipper, make sure the person helping knows enough about drysuits to draw the sliding tab with a slow, steady pull (no jerking or speed zipping), to keep underwear and the inner flap from bunching or from snagging on the teeth, to inspect for loose threads, and to not try to force the zipper past jams. (Loose threads should be cut or burned off.)

Neoprene drysuits come in two varieties: closed cell and crushed. A closed cell drysuit is made of the same material as a neoprene wetsuit. In fact, it's pretty much like an oversized wetsuit, has the same buoyancy charac-

Back mounted zippers require help in donning. The diver at right is sitting on rocks that are covered with barnacles, whose sharp shells easily cut through drysuit material of any kind.

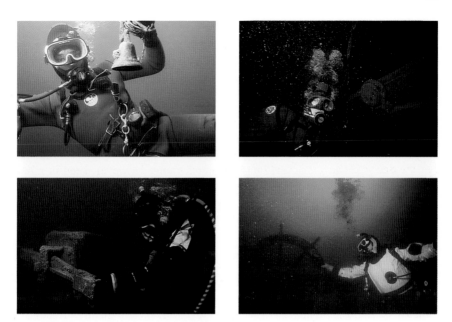

Clockwise from upper left: a closed cell neoprene drysuit, a crushed neoprene drysuit, and two types of shell suit. Of the many colors available, the most visible are red, orange, and yellow.

teristics, and offers the same amount of thermal protection. By itself, it is the warmest drysuit around and therefore requires the least amount of underwear. Drawbacks: the neoprene compresses at depth with a consequent loss of buoyancy and warmth; the quarter-inch neoprene must he offset with a massive amount of lead (I need thirty-five pounds); the thickness of the material restricts maneuverability, especially at the elbows and knees; the bulk and large surface area produce great dynamic drag: it is the worst fitting of all drysuits. On the good side, if the suit floods it maintains most of its buoyancy.

Crushed neoprene is just what its name implies: neoprene that has had the air squashed out of it. Some thermal protection is afforded by the material which, since it is already compressed, is not lost at depth. (But in any case it is not as warm as a suit made of non-crushed neoprene.) Likewise, the suit's buoyancy characteristics remain pretty much the same regardless of depth. The material stretches and flexes, making the suit comfortable both in and out of the water. Less lead is needed than with closed cell neoprene.

Vulcanized rubber and nylon fabric drysuits occupy opposite ends of the monetary spectrum, but have some common attributes: a sleek surface that promotes glide and that dries quickly, choice of neoprene or latex seals for the neck and wrists, lack of thermal protection, little need for lead, and no change in buoyancy characteristics.

At more than twice the cost, vulcanized rubber offers the elasticity that nylon fabric wants, resulting in a drysuit that is manufactured with a form fit that does not wastefully trap air. Rubber makes the drysuit heavier, but so

rugged and long lasting in such adverse conditions that commercial divers swear by them.

Nylon fabric, on the other hand, has no stretch at all. To enable a person to get into a fabric suit, and to get around with it on, the suit must be made to fit loosely. The result of this somewhat baggy appearance is the sobriquet "garbage bag suit." This unfortunate connotation demeans the product more than necessary. Although the waterproof urethane backing eventually peels off, a condition that is "whole body" and is not repairable—it deprives the material of its ability to keep out water—the suit is moderately priced, and the person who makes only a couple dozen dives per season will get many years of use out of it. Why buy a Rolls-Royce for a monthly trip to the bank?

The nylon fabric drysuit is more rightly called a "shell suit," since its purpose is to provide a waterproof covering for heat retaining undergarments. Because the material is so thin, the suit weighs practically nothing, can be rolled into a small package, and is as comfortable to wear as a rain coat and rain pants.

A compromise between vulcanized rubber and nylon fabric is a "halfway" drysuit consisting of butyl rubber backed on both sides by nylon: a material known as tri-butyl laminate. It's both lightweight and flexible, and ranges in cost from reasonable to expensive. In a pinch, a pinhole can be temporarily quick-fixed with duct tape.

Drysuits take a little getting used to. Perhaps the most important rule is: don't get upside down. In fact, be careful even when you're horizontal because, if your suit has a lot of air in it, and you tip forward, the air might rush up your legs and into your feet where the concentrated buoyancy will flip you over and send you unwillingly and ignominiously to the surface. Sometimes you can recover from an inversion before broaching like a wounded whale, but not always.

When I got my first drysuit I devised an ingenious way of carrying my goody bag where it was readily accessible: snapped to a rope looped around my neck. It was easy to shovel in lobsters and scallops, and, on occasion, small artifacts. It worked fine until the day I found a porthole. On my way back to the ascent line my two buddies also found portholes. They liked my goody bag technique so much that they decided to employ the same device—and deposited their portholes in *my* goody bag. I went up the line like a brass zeppelin, with my drysuit inflated to mammoth proportions, and me flopping around inside like a stick in a rattle. As I neared the surface the air in the drysuit expanded exponentially. Suddenly I capsized. My drysuit legs became two sausage balloons, the portholes around my throat became a neck-stretching anchor, and my arms were thrust out to the sides with the rigidity of rigor mortis. I was helpless, like a victim on the rack. I gripped the line in fear.

Eventually, my buddies noticed my predicament and came back down to help me out of my topsy-turvy position; they were afraid I would drop the bag with the portholes. The moral of the story is obvious: never trust your buddies when it comes to recovering artifacts. Also, never put your neck in a noose unless you want to hang yourself.

Since it took two strong men to right my keel, I fully understood the feeling of helplessness many years later when I saw a friend's fins bob to the surface at the ends of booties that had no feet in them. He drifted along bottom up, unable to effect his own rescue. Fortunately, he remained calm until I got

there to pull off his useless fins, flip him over, and tow him back to shore. (He could not get his feet back into the booties, and the fins were in danger of falling off.)

Another time, on a cold dive in April, one fellow had to stab his drysuit legs with a knife in order to abort an inverted sub-orbital flight. Should these didactic yarns concern you? Yes, because such situations can be avoided by learning "drysuitable" techniques.

Most overinflation problems are caused by having too much air in the drysuit, usually the effect of wearing too much lead. Balance is then lost when the head gets lower than the feet and the air shifts sudden enough to catch the diver unawares—such as when you peer under a hull plate and try to outwit a wary lobster. If you hang onto the wreck you can usually bring your legs against your chest and regain a proper attitude. It is much more difficult to right yourself when you're free-floating and shooting for the surface. The tuck-and-roll must be performed before the drysuit's arms and legs are stiffened by air to the point of inflexibility. Otherwise, the best you can do is to flare out in order to increase your drag (thus slowing your ascent), and remember to exhale. Pray that you don't come up against the bottom of the boat.

A properly weighted diver who keeps his buoyancy in control will not go ballooning. You must have enough ballast to descend freely without the use of a line, and to halt your ascent when your tanks are empty in order to float beneath the surface at an untethered safety stop. More weight than that is a liability, as is any weight you pick up during the course of a dive.

If you wear a BC in addition to your drysuit, use the BC to adjust buoyancy and keep the air in the suit to a minimum. If you don't wear a BC with your drysuit, don't overweight yourself or try to bring heavy objects to the surface. Even if you manage to remain upright, the excessive air in the suit might "burp" past the neck seal deflating more than your ego.

Do not pull down the anchor line in the typical head-down posture. Remain horizontal or slightly head up, vent air to begin the descent, and add small bursts to compensate for compression—the material squeezing against your body will tell you when to add a puff through the inflator. On the way up keep a constant check on your buoyancy and rate of ascent. Slow and easy does it.

Fin strap keepers will keep your fins from popping off, even if the fin strap breaks. Ankle weights do double duty: they distribute ballast so you don't have an excessively heavy, back-breaking weight belt, and they help maintain trim by keeping your feet lower than your center of gravity. I have found it more efficient and comfortable to wear my ankle weights just below the knees, where the lead is closer to the pivot point of the kick. (Everyone knows that in lifting a broom parallel to the floor, it's easier to pick it up in the middle than at the end of the handle.) Lead at the knees has the same trim characteristics as lead at the ankles.

A drysuit, like any other piece of gear, will suffer mishaps or mechanical failures only when in use. This means that sooner or later you're going to get wet. People who use drysuits a lot know that a more appropriate name for a drysuit is a "dampsuit." Perspiration can't escape, so it collects in the underwear. Pinholes appear magically. Lean folks with thin wrists will find that water shoots up their sleeves when they make a fist or grab something firmly, because of the channels created by flexion of the tendons: channels that no seal can conform to. Rubber bands or Velcro straps on the neoprene wrist seals can reduce or curtail such leakage. If you have latex seals, watch out for dry rot, and puncture wounds caused by jagged or untrimmed fingernails.

There is also the possibility of massive flooding. One time I thrust my hand under a steel plate for a quick-witted lobster, and tore a gaping hole in my sleeve on an unseen metal protrusion. Water crept up my arm with such celerity that I immediately terminated the dive and dashed for the anchor line. By the time I got on the boat I was drenched right down to my socks. I didn't even get the lobster.

I hung out my undies to dry, patched the suit, and was diving dry again the next day. Leaks are to diving as fender benders are to driving: no one likes them, but everyone has them. Drysuit diving has its ups and downs, but it beats going naked.

"Get it away from me! I was kidding when I said I wanted crab cakes."
(From *Twenty Thousand Leagues Under the Sea*, circa 1872.)

From Shore to Shipwreck

No pensive person who has lain on the beach on a lazy afternoon, soaking in the sand and sun and watching children gambol in the nearby shallows, can be unmoved by the mystique and mystery presented by the sea. What underwater treasures lie beyond the breaking surf? What unknown adventures await? The lure of that hidden world, forever veiled in its deep aqueous cloak, must exert a pull on every questing soul.

I was but a prepubescent lad when I bought my first face mask and dived into the briny deep to view the submerged wonders of creation. It was a short peek, however, for Mother Nature played me a cruel trick. She scooped me up with a monstrous curl of water, tumbled me about until I did not know which way was up, then ungently deposited me in the wake of fast receding waves. Hand over hand I crawled up onto the beach, gasping, while the undertow took my mask the other way. She was not yet ready to let me gaze upon her beauty.

A decade later, by that curious quirk of selective memory, I had forgotten that long ago episode with the pounding surf. I donned another mask, along with a tank and regulator, inflatable vest, wetsuit and weight belt, fins and snorkel. My goal was only a hundred yards away, the sunken yet visible concrete ship *Atlantus*, off Cape May, New Jersey. I was only halfway there when Mother Nature said, "Don't you ever learn?" and promptly swept me out to sea on an ebbing tide.

My imagination was assailed with horrible visions of drifting helplessly away from land, of becoming lost in the broad ocean swells, of spending a dreadful night afloat, of ignominious rescue—assuming, of course, that my buddy's wife would report to the Coast Guard the direction we had taken. Through supreme effort of will we caught the last jetty marking the end of the Delaware Bay. Again I crawled up onto the sandy beach, gasping. Mother Nature was kinder this time—she took only my pride. I have since learned about tide tables.

Quite a few ships have piled up on the rocks and shoals extending from the coast, and nothing seems simpler than slipping under splashing holiday frolickers to visit these crumbling remains a few hundred feet offshore. While surfcasters struggle to reach the placid water on the other side of the breakers, you could be surrepticiously spearing fish or observing shipboard structural components in relative solitude. Despite this presumption of ease and accessibility, diving wrecks by swimming to them from shore is not as convenient as you might think.

A dive may be stymied before you even don your gear, for, although we live in the land of the free, many littoral areas are controlled by the government (local, state, or federal) and some are posted private property. This does

The concrete ship *Atlantus*. Left: Joe Thompson maneuvers frantically to avoid being impaled on steel reinforcing rods that stick up like punji stakes. Right: I found easier access a few years later, when the Delaware Bay froze solid.

not necessarily imply that the wrecks are off limits, only that access from the beach is forbidden; in most cases you are permitted to dive a wreck from a boat. Before you chance a fling with the law or traipse across someone's lawn, secure the proper permission.

Then there is the problem of dressing for the dive. If the site is off some lonely strand or in a remote wilderness location, your only contention may be the weather. Not that stripping down to the nude in the wind or freezing rain is any bargain. But you do need to be mindful of people in the neighborhood whose moral codes are offended by pre-dive briefs. Like that time in the province of Nova Scotia when one provincial observer, safely ensconced in a distant home that was perched high on a bluff, gawked at me and my female

The way we like to imagine shore dives. Turn the page for a reality check.

The dream is over. Eric Garay shows how it's really done, by climbing down a boulder face with doubles on his back. At right, he leads the way down a densely overgrown hillside.

companion until the amount of skin exposed exceded the voyeur's threshold of respectability. What should next appear on the horizon but the constabulary in the guise of the Royal Canadian Mounted Police—in a car, not riding a horse like Dudley Do-Right. Some quick explanations and an officer who was reasonable kept us from doing our surface interval in the slammer. Remember that the world is populated by busy-bodies who would rather mind your business than their own. Watch out for them. In the long run, you'll turn fewer heads by squirming into your thermal suit in the back seat of your compact.

Worse yet is the often monumental task of getting to an entry point that is close to the site.

Most ships are inconsiderate enough to come to grief in places that are far from the maddening crowd. Unless condos and commercialism have over-taken the area, these wrecks might exist today in dire straits on distant banks; they require research to locate and ingenuity to reach. Driving along a deserted beach in a four-wheel-drive vehicle is fun even if you never get to dive. Park well above the high tide mark. I once saw a fellow four-by-four that had stalled out at the water's edge when the tide was out; it was still there when the tide came back in. How do you explain to your insurance company that you sank your truck?

Sometimes you can't approach the entry point with any kind of wheeled vehicle. This means walking. In extreme cases it means hiking or climbing. The rocky coastlines of Maine and Nova Scotia come painfully to mind. Trailblazing through forests, descending steep cliff faces, and scaling boulders all require strength, stamina, and a will to survive.

Furthermore, you've got to carry heavy equipment that gains in weight with each successive step. If the distance is not appalling and the temperature less than stifling, you might opt to don your suit and tank and carry the rest of your gear in your hands. That's about as much as the average person can manage on a path that is smooth and level or on a short stretch of beach. If the terrain is rugged or hilly and the distance fair to middling, you should consider transporting your gear in stages. This is done in two ways.

The short form is to make one trip carrying your hardware (tank, regulator, and weight belt) on your back, and a second trip garbed in your softwear (wetsuit or drysuit) with your miscellaneous accessories in hand. This can only be done if you can stand the heat that is generated by the exertion and contained by the rubber.

The long form is to make the second trip carrying your gear bag with your accessories (including lights, camera, and action), then make a third trip with your softwear: either wearing it or carrying it.

If you decide to dress on site, consider wearing a bathing suit under your clothes, for protection from the iniquities of inquisitive onlookers. If you elect to wear your drysuit for the journey, galoshes will go a long way toward ensuring watertight integrity. And remember that carrying a drysuit also means carrying all the underwear that goes with it.

If you've got to scramble over rocks, squish through marshland, navigate around trees and brush, or go a long way in soft sand, sturdy shoes or hiking boots are a must. Neoprene booties were not designed for outdoor trekking; your feet will slide around inside, putting you at risk of tripping, slipping, or spraining an ankle. Pebbles and sharp sticks will be more discomforting than the pea was to the princess, and will poke through the soles of your booties with disconcerting ease. In this case, galoshes are inadequate because they don't offer ankle support.

Gear bags are decidedly uncomfortable for the long haul They are awkward to carry either by your side or over your shoulder: thin straps bite into your hands, and the unequal distribution of weight can throw you off balance while negotiating narrow ledges. I carried my gear in a backpack long before it became fashionable. Nowadays, combination backpack/gear bags have entered the retail market. For anyone who wants to pursue shore diving assiduously, they are the way to go. Thick, padded straps ease the strain on the shoulders, a cushion protects the back from metal or hard plastic equipment, and the material is washable.

Some shipwreck sites are so remote that you have to supply your own

You can't get much closer to land. The wreck on the left is in Bermuda; the one on the right in Nova Scotia.

air. I've dived in the Canadian provinces where the nearest town with a scuba shop or air filling station was hundreds of miles away. Either you haul lots of tanks or bring your own compressor. A portable compressor is a fair-sized investment unless you're serious about diving wrecks that are far from the beaten track. Yet the experience can be rewarding if only because so few people have been there.

The *Columbus* is a case in point. Ensconced in a hidden cove on the north shore of Lake Superior, a single dive in twenty feet of cold, clear water required driving off the paved road along twenty miles of dirt, fording several small creeks, then, because of the number of trips required to pack in all the gear, hiking trails a total of eight *miles*. It took all day.

Efficiency counts on this kind of dive. Pre-rig as much of your gear as possible. Put your regulator on your tank and make sure the tank is full. If you wear a wetsuit, strap on your knife. If you carry your drysuit, take talc or silicone spray for the seals. Make a check list to ensure that you don't leave an essential item behind. And most important of all, don't lock your keys in the car.

Let's assume that you've made it to the entry point with all your gear, and still have the energy to make the dive. It's time to get dressed. Take special precautions if you're on a sandy beach. Lay out a blanket or plastic tarp for a dressing station. Be careful to keep sand out of your booties and regulator. After you've taken your gear out of the bag, put your clothes inside in case of rain. If the beach is crowded with kids, obnoxious teenagers, or adult unsavory types, set land mines around your perimeter. Then shuffle down to the water's edge with your mask around your neck and your fins in hand.

Walk straight on in. Lean against your buddy while you pull on your fins, or sit down in the shallows if you're independent. Once those clodhoppers are on your feet you're as klutzy as a two year old; worse with the water

The trail to the *Columbus* is long and the footing insecure, such as when crossing this old bridge, but the dive is well worth the effort. the *Columbus* is a wooden steam tug that sank in 1910.

Left: A portable compressor may be necessary for wilderness wreck diving. Right: surface interval on the rocks. Drysuit undies are laid out to dry, empty tanks are exchanged for full ones, and there's time for lunch and relaxation.

eddying around you or with waves hitting you broadside. Seat your mask, take a compass bearing on land, and go: snorkeling if you need land ranges to locate the wreck, underwater to avoid a heavy surf—you can always pop back up on the other side of the breakers to check your bearings. Do not surf the breakers on your return; coming down off a charging crest with a tank on your back is not the way to fly. Stay on the bottom and grip the sand with your fingers, both kicking and pulling against the undertow until you wash up in shallow water. Depending upon conditions, either crawl out of the water on your hands and knees, or remain sitting until you get your fins off, then stand up and walk ashore.

If you feel awkward in the transmigration zone—when you change from a land animal to a creature of the sea, and back again—temper your feelings with the recognition that every shore diver suffers the same lack of aplomb. You are in your element on either side of the interface.

Diving from a rock-bound coast presents entry and exit problems of a different nature. A craggy, boulder-strewn jump-off could be more of a military exercise than a leisure time activity. Rock surfaces at the water's edge are often made hazardously slippery by algae (fresh water) and seaweed and kelp (salt water). Vagrant swells or the wakes of passing vessels may slam you up against submerged pinnacles before you can get away from the shallows and into deeper water. On the other hand, tidal swells are periodic and can facilitate entry. I can remember sitting on a kelp bed at low tide, sidling closer and closer to the edge until the incoming surge gently lifted me up and floated me out to sea. I used the same water cycle to get back up on land, except that the slick kelp was then working against me. I squirmed and scrabbled up the kelp-covered rocks with the help of the surge, then, when I didn't make it all the way, was left high and dry during the outflow and slid helplessly back into the water. This kind of roughhousing is tough on the gear, drysuits, and especially on cameras.

You must also learn to anticipate the amount of tidal change that will occur while you're off exploring the wreck. Too many times I've returned to my point of embarkation only to find that the water level had dropped and that the rock ledge I had used to stage my entry was several feet overhead. A rising tide can cause similar problems. It pays to scout the surrounding area and note alternative exit spots. Once I had to use a rope to climb up out of the water. If you've got a choice, know that the best window is during slack tide, either high or low. Decide how long a dive you want to make; if you go in half that amount of time before dead slack, you'll be coming out the same amount of time after dead slack, and theoretically at the same water level.

River entries possess a peril all their own, not the least of which is a stiff current that never lets up. Either set up a shuttle or keep cab fare in your pocket, since you may wind up so far downstream that the walk back would cripple you. River divers often establish permanent guide ropes from shore to shipwreck, and use them as static lines. They swim to the wreck clipped off with a carabiner, like an arctic explorer leaving an ice cave in a pea-soup fog, then follow it back again.

If you slip off or lose the line or get carried away by the current, swim toward shore. "Do you think me daft?" you are probably saying. "Anyone with the brain of a potato knows that." But you'd be surprised how impulse drives people to buck the current in an obtuse attempt to reach not just any point on shore, but the point from which the dive began. When I say to swim toward shore I mean it both literally and littorally. *Straight* toward shore. It takes no longer to make landfall in a five knot current than it takes in no current. The difference is a vector product: your flight path describes a diagonal that is the combination of your magnitude and direction. You'll just end up a bit downcurrent, as I did when I got swept away from the *Atlantus*.

Now that we've overcome access, let's deal with traffic. Curiosity killed the cat as well as the impatient diver. If you hear engines overhead, *stay down!* Let the air out of your BC or drysuit. Hug the bottom. Lie low till the sound goes away. Chances are, the ship or boat wasn't that close anyhow—it just seemed that way because the speed and volume of sound are augmented by the density of water. But there's no sense taking a chance of getting run down by an inbound steamship or minced by an outboard motor. Wait a sufficient time before coming up.

Then ask yourself why you didn't have that marker buoy along like you were supposed to. Seal teams and UDT divers don't want anyone to know where they are, but wreck diving is not a clandestine operation. Not only will a marker buoy warn boaters of your presence, it will tell your friends that you are on route to the wreck, have finally made it, are on the way back, or are headed out to sea.

In its simplest form the marker buoy is a float with a diver-down flag mounted on a short staff, and a line by which to tow it. The float can be inflatable or solid styrofoam, the staff can be wood or plastic and can have a counterweight slung below the waterline, and the flag can be fixed or free-waving. It's okay to tie it off on the wreck once you get there; its purpose is to let people know that a diver is down and about, not to pinpoint his position with three decimal accuracy.

A larger variety consists of an automobile inner tube fitted with mesh. Not only is it easier to spot, but it does double duty as a topside goody bag. You can use it to carry extra equipment, store fish that you've shot or seashells you've collected, and store snacks and thirst quenchers for those super long dives: a tank goes a long way in the near shore shallows. You can leave it there for the day; it will mark the site for your buddies and make the wreck easier to find on your next dive.

Local ordinances often mandate the use of a marker buoy, but even if not it's a good idea.

Upper left: Jim Aitkenhead hands equipment to divers who have made the transfer from land to water. Lower left: Rick Jaszyn rests after a difficult sturggle up the slippery kelp bed. Below: Steve Sokoloff tests the surge and crashing breakers before donning his tanks—a kind of "wet" run before making the actual entry and subsequent exit.

The Boat: Rite of Passage

I hate boats. They are either too hot or too cold, always cramped, uncomfortable, and upsetting to my vestibular nerve, and if I never saw one again in my life, it would be too soon.

Strange sentiments, you might say, from one who in two decades of diving has accumulated more than three years worth of days at sea. Nevertheless, that's exactly how I feel. To me, boats are a necessary evil I must endure in order to reach the wrecks—the exploration of which I love. No born follower of the sea am I, nor enticed am I by the Sirens to visit distant foreign lands.

Having said my peace, I will now embark upon a journey of vicarious enlightenment from my temperature controlled, spacious, and immovable study, and relate the joys and jeopardies of diving from boats. As seen in the previous chapter, diving from shore has quirks that cannot always be overcome by sheer strength and stamina. Ships that sank too far away from land, or off illegal entry points or unclimbable geologic outcrops, must be gained in different ways. This means boats.

I've never heard of anyone diving from a canoe. I have no doubt it's been tried, but I think the vessel's inherent instability precludes effective use. A small, flat-bottomed scow and a pair of oars may be practical under special circumstances: close to shore, in flat water, and with no wind. And solo dive kayaks are now being marketed. Generally, though, the minimum type vessel suitable for diving is the inflatable boat.

I'm not talking about a plastic life raft or kiddy creek runner, but a full-sized inflatable constructed of strong synthetic or vulcanized rubber. When most people think of inflatables they think of Zodiacs, a trademark that has become as synonymous with rubber boats as Kleenex with tissues, Band-Aids with adhesive pads, and Exxon with oil spills. With the modern day penchant for the abbreviation of multisyllabic words, the generic derivative "zodes" has evolved to mean inflatable boats of all kinds.

Since this book is a user's guide, not a buyer's guide, I'm not going to describe all the features that differentiate one model from another. At the low end of the spectrum are the affordable brands without the costly extras, while in the five- and six-figure range there are boats so tough that the Coast Guard employs them as rescue craft, and the military has some made out of bullet-proof kevlar which are used for riverine operations in hostile territory. (Assuming that wreck divers do not become the targets of regulatory execution, kevlar is an option we can live without.)

I do, however, want to impart enough descriptive detail to instill confidence in the product. Zodes are incredibly stable and nearly impossible to capsize. Even if totally flooded they will remain afloat. Collapsible floor boards can be installed to give them a semblance of semi-rigidity and to make a solid platform for footing and gear stowage. The thick-walled pontoons are extremely resistant to puncture wounds—a fact that is not intended to imply that you can stab it with an ice pick or be flamboyant with your dive knife. In truth, the least dependable part of an inflatable boat is the motor you procure to propel it through the water.

Nothing is more frustrating than pulling a starting cord over and over and over, until your arms hang limp at your sides, only to have the motor cough and sputter and defy all attempts to start. It's more than frustrating when you're several miles at sea, facing the prospect of rowing all the way to shore. Get a good motor, keep it well maintained, understand the rudiments of operation, keep the essential spare parts on board, and don't run out of gasoline: AAA won't be able to help.

Zodes are great for accessing shipwrecks normally thought of as shore dives. Long walks and difficult entries are replaced by high-speed dashes out of protected coves. However, there's still a great deal of work involved in using a zode, such as assembly and transportation. Unpacking and inflation on site can take a couple hours; if you keep it pre-assembled, you'll need a trailer to get it to the launch ramp. If no ramps are in the vicinity, you'll have to carry it over sand, mud, or rocks, then you'll have to do the same with all your gear. Zodes are heavy even when empty; it usually takes three or four people to lift one.

Zodes are wet; fast zodes are wetter. Whenever I've dived off of zodes I've donned my thermal protection on land. This usually means wearing it all day. On a boat without shade in the heat of summer a person can get mighty hot—especially the one who has to stay topside while the others make the dive.

Which brings us to the cardinal rule of boat diving: *never leave the boat unattended*. Never! I've heard too many horror stories of boats breaking free

Zodes are heavy even without the motor installed. They are also cramped once they are loaded with dive gear.

while all the divers were on the bottom. With the gunwale acting as a sail, the wind can blow a boat away a lot faster than most people can swim—even if they ditch all their gear. Then there was the idiot who abided by the rules without using common sense. He left his girlfriend on board while he went diving, but never told her how to start the motor or operate the radio. The anchor line broke, the boat went adrift, and she stood helplessly on deck for hours until she was able to wave down a passing vessel. Don't add a false sense of security to your list of sins.

The main advantage of a zode—or any small boat, for that matter—is mobility. The fact that it can be hauled to remote dive sites means that you can reach wrecks previously unaccessible to you, or that were difficult to access. You will still have to rely on ranges to locate the wreck (or marker buoys if they have been established), but from the vantage point of height you've got a better view of the landmarks, you can move around quickly and with little expenditure of energy, you can deal effectively with situations caused by current, tide, and sudden changes in the weather, and you can save the air in your tanks for diving the wreck instead of fighting through a heavy surf and saving enough air to make the long swim back. Plus you've got a diver-down marker float the size of—well, a small boat.

If you're not quite sure where the wreck is, send a diver down on the anchor line to conduct a search. He can either carry the grapnel with him and tow the boat along (or topside personnel can feed out line), he can swim about towing a wreck marker (usually a bleach bottle on a string), or he can tie off a line reel and make sweeps (this technique will be covered in the next chapter). Once the wreck is located, the others just roll over the side and follow the line down to the bottom.

Donning gear on a zode takes practice, balance, and not a little help. Until you've dived off one, you can't believe how jam-packed it can be when divers, tanks, and gear bags are all competing for the limited square footage. Depending upon the number of people, there may hardly be enough room to sit, much less to move about and get dressed.

The simplest way to get into a tank harness is to place the tank on the pontoon, have someone hold it for you, then crouch onto your haunches and slip on the shoulder straps. *Do not cinch down the straps or fasten the waist belt until your regulator is in your mouth and your BC or drysuit inflator is connected.* The position on the pontoon is precarious at best, and an errant ripple can make you lose your equilibrium and fall backward off the boat. Besides, at this point your butt is hanging in the air. Ease back and sit securely on the pontoon, lean inboard, and put on the rest of your gear. The most secure method for donning tanks, especially when the boat is bobbing like a cork, is to sit on the floor and sidle back into the harness, then have someone pull you up and help you to a sitting position. But don't ever sit on the pontoon unless you or your tanks are being held. Once all your gear is on and your BC is inflated, simply roll over backwards. Bon voyage!

Another method—one that does not require help—is to pull on your fins

They spray soaks everything and everyone, so we usually don thermal protection for the trip to the dive site. Sometimes, due to the local geography, zodes must be launched miles away from the wreck and then be driven along the coast. You can see the buoys marking opposite ends of the wreck. In this case, we unloaded extra tanks and gear on shore, then shuttled people out to the down lines.

Lonely vigil; a diver waits for his companions to return from the bottom. Sometimes entire flotillas of zodes get together for mutual safety. The work is never done till the zode is placed back on the trailer. During surface interval, a group of divers motors behind a pod of porpoises.

first, slide back into the harness, connect your BC or drysuit, then roll over onto your knees and complete dressing from there. To get into the water you can either bend over the pontoon and slide in head first, or you can roll over the pontoon sideways.

After you've successfully gotten off the boat, you've got to get back on. Hand up any loose equipment (lights, mesh bags, etc.) to the topside sentinel, or clip it off to a line hanging in the water for just that purpose. These items are lightweight and seldom cause a problem, but more than one weight belt has been dropped during the transfer. Plan ahead by weaving a D-ring onto your weight belt; then you can clip the equipment line snap-hook onto it, release the buckle, and pull the belt free from your waist and dangling hoses. Either let it be hauled up, or ease it down gently; a sudden drop could tear out the grommet to which the line is attached.

Spin around so your tank faces the zode, let your buddy wrap an equipment line around the valve, then release the waist buckle, shrug out of the harness straps, and get away from the tank. If you have a BC that is built into the backpack, inflate it fully before you take off the tank.

Grab onto the pontoon, kick hard with your fins and pull with your arms, and over you go. Topside personnel can help by grabbing your armpits and hauling you in like a sack of potatoes. You'll undoubtedly fall into the bottom of the boat with the dignity of a newborn seal. It's okay to laugh. Perhaps it's even necessary.

While zodes can get you to nearby wrecks up to several miles away, for better comfort and longer range you'll want a more seaworthy craft. This generally means a power boat: one with a cuddy cabin or a center console and spray skirt. A bigger boat means more available space for gear stowage and dressing, larger fuel capacity, and a vehicle that can be outfitted with wreck-finding electronics such as a depth recorder and loran. As with zodes, it is not my purpose to advise you how to buy or operate a boat, only how to rig it for diving if you're new to the game. Your bank account will determine how much boat you can afford, a dealer will try to talk you into spending more. The Coast Guard offers courses in small boat handling and "rules of the road," although the rules actually apply to the sea and to inland navigable waterways.

It is also beyond the scope of this book to teach you how to use sophisticated electronics. Instruction manuals come with each piece of equipment, although you'll find that experience and continued use will be your best training aid. But if you already own a boat (a fisher turned diver), or if you have the opportunity to dive off the boat of a friend who does not understand the specific needs of wreck diving, or if you would like to own a boat yourself someday so you can dive independent of the group charter services, then it behooves you to learn the rudiments of grappling, establishing safety lines, and safe entries and exits. This knowledge will also enable you to educate charter boat captains who specialize in fishing charters but who occasionally take out divers, and who might not be familiar with wreck diving procedures.

Left: Cris Kohl scratches his head and Don Edwards holds out his hands as if to say, "How are we going to fit all this stuff on the boat?" In addition to dive gear we've got food and camping equipment for four people, plus a portable compressor (see photo on page 63). We spent a week diving the wrecks off Isle Royale. Right: Notice the makeshift ladder on this fishing boat I chartered in Nova Scotia. It consists of floor boards nailed across two sapling trunks, and is tied with twine to the roof.

If you're diving a wreck on a shallow tropical reef, where the visibility ranges between great and awesome, chances are you can anchor the boat in the sand within a hundred feet of the wreck, and you'll see it when you jump in— maybe even before. But in the cooler waters of the north, where wrecks tend to be deeper and darker, you must set a direct line from the boat to the wreck, else the divers may never find it. This is sometimes done by dropping an anchor in the sand a couple hundred feet upcurrent of the wreck, then drifting back and snubbing off the anchor line when the spike appears on the depth recorder, and lowering a weighted line into the wreckage or debris field, to be used for descent and ascent.

It is more common, though, to drift back with a grapnel which drags into the wreckage till it hooks. When the current or wind pulls the boat around and the anchor line becomes taut, the "hook" is set. In this case the anchor line does double duty: it holds the boat on site and acts as a descent/ascent line. The biggest mistake made by small boaters is in using a grapnel that is appropriately sized for the boat but not big enough to do the job. A grapnel that is too light will not fall fast enough; it will sail over the wreck if the boat is moving rapidly due to high wind or current. One trick to overcome this problem (other than the obvious—getting a bigger grapnel) is to use a long, thick chain as a leader. Since a chain is necessary anyway—a rope is soon cut or chafed through on sharp metal edges or encrustation—it seems the best solution; you can shackle on additional lengths until you have enough weight. However, if the chain is too heavy it will fall faster than the grapnel, creating a loop in the line; then the grapnel may snag on its own chain instead of on the wreck. A weight attached to the nose of the grapnel may be a better option. Whatever method you choose, I guarantee that you will get more than enough exercise by pulling up grapnels that miss. When choosing or making a grapnel, keep in mind the fact that if the tines are insufficiently strong (that is, too slender) they may straighten out in a strong wind or current or when the boat is lifted suddenly on top of a large swell.

Releasing a grapnel requires a little finesse which does not come easily. The quickest way is for the last diver to pull it free from the wreck. While this sounds simple, there are a few things to consider when doing so. First of all, make sure you're the last diver in the water; it is an unpleasant feeling to return to the anchor line only to find it gone. Prior communication on the surface will ensure that no one gets left behind. If you've established a sequential protocol stick to the program and don't dive out of turn or you may quite literally miss the boat.

The actual "pull" can be a ticklish job if the waves or swells are high or the current strong. When the boat rises and falls, the anchor line alternately tightens and slackens. In the textbook scenario you grab the grapnel by the unhooked tines, wait for slack in the line, pull back hard and fast, and toss the hook aside or over the snag before the tension returns. The actual performance is subject to numerous variables and seldom works with such precision.

If the slack period is not long enough there may not be time to unhook the grapnel before the anchor line tightens. This occurs when the tines have gouged into or become twisted in the wreckage, or when the chain has sliced a groove in the metal and the links become wedged. Extrication requires good timing and strong man-handling. *Never* get downcurrent of the grapnel when releasing it or "working" it loose; if you get snagged you may get keelhauled along the wreck, pinned against other wreckage, or dragged completely off the site, all of which invite injury or damage to your equipment. In short, pull the grapnel—never push it.

It is important to make sure that you don't get your fingers pinched and that your hoses are unentangled.

Assuming that the grapnel comes free, there's always the chance it will rehook itself before it drags clear of the wreck. One way to avoid this potential problem is to "foul hook" the grapnel. To do this you flip the grapnel back on the chain, then loop the chain around a couple of tines that are opposite each other; this way, as the grapnel drags across the bottom the tines are pointed away from the direction of travel and are not likely to snag. Nevertheless, instead of ascending the anchor line immediately, hover within sight of the grapnel until it passes beyond the perimeter of the wreck: even a backward grapnel can snag if it lodges between beams or falls into a hole in the hull.

Another method of keeping the grapnel free is called "riding the hook." It's great fun, but should not be attempted until you have acquired enough experience to feel comfortable with your dive gear and at ease with pulling the grapnel. Then, after you get the hook free you hang onto the tines and let the drift of the boat take you in tow. Make yourself positively buoyant enough to offset the weight of the grapnel, so you won't hit any high points, then lean back and enjoy the ride. It's like a flyby sightseeing trip. I usually stay down until the grapnel is well away from the wreck, and look for attendant structure that may never have been seen before.

If no provision has been made to pull the hook, it has to be jigged from the surface. This takes practice and patience. After all divers are out of the water and the appointed deck hand is stationed on the bow, the operator puts the engine in gear and moves slowly ahead while the deck hand pulls in the slack in the anchor line. Careful coordination between operator and deck hand is necessary. If the boat moves forward faster than the deck hand can haul in the rope (now wet and slippery), a loop results, and this loop may foul the boat's propeller. Once the deck hand signals that the anchor line is vertical, the engine is idled and the boat is allowed to drift slightly forward while the deck hand "jigs" the grapnel. "Jigging" means alternately heaving and dropping the line until the tines unhook from whatever they were caught in. Then the rope is hauled up rapidly until the grapnel is clear of the wreck, whereupon a slower pace can be maintained.

Jigging can be frustrating because grapnels have a nasty habit of doing their job too well, and of rehooking on nearby wreckage once pulled free of the original snag. Instead of coming loose, a grapnel sometimes tightens its grip on the wreck. This means that a diver must be sent down to pull the hook. It is not safe to "run" the boat while divers are in the water; an uncontrolled ascent by the diver or a misjudgment on the part of the operator may result in the diver getting run over. Instead, the diver descends the anchor line to the bottom, waits for the deck hand to feed him slack, then works quickly to release the grapnel and carry it off the edge of the wreck. Careful timing is essential. If slack is fed to early, the diver may not be able to get in position before the boat runs out of rope. It's better for the diver to have to wait for slack: that is, to overestimate the amount of time it will take him to reach the bottom depending upon the depth and current. Usually, a predetermined time is set so both the diver and deck hand will be ready to act.

Since it is general practice for people to undress soon after emerging

Below: A grapnel in the classic position; notice the sisal rope at the bottom of the shank, tying the hook to the wreck. Right: This grapnel is so heavy that, instead of having a deck hand haul its deadweight to the surface, a diver is sent down to attach a liftbag on the top ring so it can be floated up.

from a dive, the person normally asked to go back down is the last one out of the water. Yet he is the last one who should be selected because he has the shortest amount of surface interval and is therefore the one most saturated with nitrogen. Someone else should volunteer or be chosen to pull the hook. In any event, a safety stop after a repetitive bounce dive is mandatory.

In the early days of diving it was customary to hit the water as soon as the engine was turned off: the silence (or the low-oil alarm on diesel engines) was the signal that the starting gate was open for those vying for the easiest lobsters and largest fish. This may still hold true under prime conditions, but usually a boat captain likes to set up auxiliary lines that make it easier for divers to get down on days when the current is working against him. Fin kicking is a wastefully inefficient means of propulsion. Even a diver in good physical shape and aerobically fit may be unable to reach the anchor line or, if he makes it, may burn up a large portion of his air supply getting there. What can turn an aborted or stressful dive into a successful undertaking is a rope running the length of the boat, permitting a diver to pull himself hand over hand to the anchor line where the descent begins.

This rope is called by several names: bow line, swim line, current line, granny line, and geriatric line or "gerry" line. In its simplest form it is tied off to the forwardmost cleat and allowed to drift back past the point of entry. Because it may slip under the boat and out of sight from above, the bitter end is sometimes tied to the sternmost cleat; keep the rope taut or it may wrap around the propeller. Polypropylene is no good because it's too slippery; manila kinks and rots too quickly. Use a thick nylon rope that a person can get a grip on.

It pays to take the time to get rid of the gap between the point where the granny line arcs up out of the water and the point where the anchor line, angling forward of the bow, comes within reach. Although only a few feet apart, the transfer from one rope to another can be insurmountable, especially on a wildly heaving sea when a good grip on either rope at those points can lift

Left: Due to an exceptionally strong current, a diver grabs the granny line before he enters the water. Middle: One diver holds onto the trail line and stays well clear of the ladder while his buddy climbs onto the boat; notice the orange float at the end of the line. Right: A side roll entry, with hoses and accessories tucked in front and one hand holding the mask.

you completely out of the water or pull your arm out of its socket. To close this gap, knot the gerry line around the anchor line and shove it down with a boat hook as close as you can to the surface, or beneath it if possible. If the knot slips as divers put their weight against the rope, you'll have to secure it tightly within reach of the bowsprit, then let out anchor line until the knot submerges. Some captains tie the granny line to a huge heavy shackle and let it slide down the anchor line of its own accord; dive weights can be added to make the shackle heavier.

If the current is severe, it is not uncommon to pull the granny line up onto the boat and hand it to the diver about to make his entry, so he does not get swept away before he can grab the line in the water.

The most difficult place to swim against the current is on the surface; not only does the diver have to buck waves breaking over him, but the wind may drive the surface water a bit faster than the water is moving only a few feet below. To overcome this interface effect, some captains lower a drop weight off the stern of the boat with the bitter end of the granny line tied off at about ten feet. When the diver enters the water he can drift back to the vertical line, descend, then pull himself up the granny line to where it connects with the anchor line at an equivalent or deeper depth. The transfer line is also used for the return; it keeps the diver at depth and away from the side of the boat which, in rough seas, may come down on top of him.

Thus the various lines (drop, granny, and anchor) serve as a single continuous guideline from the boat to the wreck and back again.

Another important safety rope is the trail line, also called the catch line or tag line. This is a polypropylene rope that is attached to a large float which is deployed off the stern. It serves a dual purpose. For divers who do not ascend the anchor line, and who are then swept downcurrent faster than they can swim to or vector toward the boat, the trail line is an extension of the boat and an aid in self-rescue. If the diver reaches the line exhausted and out of breath, he can either rest until he regains the strength to pull himself to the boat, or he can signal for help, in which case the crew can haul him in. The trail line also serves as a place to hang out while another diver climbs the ladder. On large charter boats the trail line may sometimes have several divers floating in line, so to speak, waiting their turn. Polypropylene is used because it floats; nylon sinks, so unless the line is held taut by the force of the current against the float, the rope submerges out of sight and out of reach.

Protocol calls for a returning diver to take up station at the end of the line unless he is severely stressed, in which case those in the water should let him butt in and lend assistance if needed. Under no circumstances should you get under a diver climbing up a ladder, or be close enough that he could fall on you should he lose his balance or grip.

In addition to divers access lines a captain may decide to suspend equipment lines with snap hooks, a spare tank for emergencies, and a hang line or drop bar (usually a weighted aluminum beam or a lead-filled PVC pipe held horizontal by ropes at either end) for divers to hold onto while doing their safety stops. As the boat rolls gently in the swells, all these ropes take on the appearance of a torn spider web wafting in a springtime breeze. Deploying all these lines requires quite a bit of time, and having so many lines in the water effectively immobilizes the boat until they can be retrieved. This means that if a diver surfaces in trouble or if he is unable to reach the trail line, the boat cannot simply pull the hook and go after him.

For this situation there is another line: the rescue line. This is several hundred feet of polypropylene with a loop on the end; for fast deployment it should be rolled neatly on a reel or carefully folded in a bucket. It is used to to send a snorkeler after a diver in distress, or to retrieve liftbags.

Enter the water from a small boat the same as you would from a zode. However, because of cleats and railings and an overlapping gunwale—all of which can snag fins, goody bags, gauges, and hoses—care must be taken when going over the side. When doing a back roll, tuck your accessories and hoses in your lap. When putting your knee on the gunwale to do a side roll, place your accessories and gauge panel outside the boat.

Getting back on a small boat can be difficult. The zode technique of doffing gear in the water and tying it off or handing it up works only for equipment; the gunwale is too high to climb over using a power kick and arm pull, even with help. A small boat may have an aluminum accomodation ladder that is intended for swimmers, but it may not be strong enough to support a fully geared diver: the rungs may bend under the weight, or the structural members may twist out of shape. Once, as I reached the upper rungs where my weight was no longer supported by the water, the bolts holding the ladder to the transom ripped through the fiber glass. I crashed into the water backwards and sank like a rock, still clinging upside down to the ladder. I saved the ladder by inflating my drysuit and bringing it back to the boat. If you have a weak ladder, you should make provision for doffing tanks in the water.

Some boat captains and training agencies recommend removing fins before climbing a boat ladder. In general, I disagree. If a ladder is so poorly

Getting onto a boat without a ladder is difficult, but can be accomplished with teamwork. First you hand up your weight belt and accessories, then your tanks are either held for you or tied off with a stout rope while you shrug out of the harness. Go on to the next page.

Now you either pull hard and kick like crazy, or let your
buddies lend some helping hands.

designed that it makes it nearly impossible for divers to get into the boat with
fins on their feet, don't make excuses—get another ladder. If you cannot climb
a ladder due to physical debility or because of particularly rough water, then I
will concede the point. But make sure that someone is watching you and that
you have a good grip on the ladder or have a safety line in hand. Once you go
adrift without fins, you are helpless against even the mildest current.
Otherwise, the safest and most common practice in wreck diving is to keep
your fins on till you are securely aboard the boat.

If you decide to outfit your boat with a ladder there are a few features to
keep in mind. A dive ladder needs to be pitched at an angle so the diver is not
cantilevered backward by the weight of his tanks. Twenty degrees from the
vertical is about right; any more, and the diver has to crawl onto the boat.
Weld or bolt a bracket to a straight ladder to keep the bottom away from the
hull.

The ideal ladder should be wide enough so that a diver has plenty of
room to twist his feet with his fins on as he steps from rung to rung. The rungs
should be spaced close enough so a diver who is undertall or whose
movements are restricted by a thick neoprene wetsuit or the bulk of drysuit
underwear can bend his legs enough to reach the next rung without undue
strain. If the rungs are round instead of flat, they should be fat enough to give
broad support to the arch of the foot. A person cannot stand on a skinny rung
in full gear any more than you can hold a heavy package with a piece of string.
If the ladder has railings, they should not be so far apart that a diver cannot
hold comfortably onto both of them at the same time. It is also important that
the ladder be long enough to extend at least four feet below the surface.

An alternative to the traditional ladder is one whose structural support is
provided by a center post and whose rungs are open at the ends. This kind of
ladder is the easiest to climb with fins on.

There are four times during an exit when you have to be exceptionally careful: approaching the ladder, getting on the ladder, staying on the ladder, and getting off the ladder. This statement may seem like droll exaggeration to one whose image of dive boat activity is based upon resort advertisements, but I am quite serious. Take for example the time I was gearing up on a boat while another diver in the water was returning from the wreck; the waves were so big that when the boat sank into a trough and the diver alongside rose up on a crest, I found myself looking *up* at him. My assessment of the situation was that getting back on the boat was going to be a singular challenge. It was.

When the seas are bad the boat will pitch and roll with charming irregularity. Correspondingly, one moment the ladder might be completely out of the water, while the next it is crashing down on anything in its way. When the stern rises high on a crest a diver in the preceding trough is likely to get sucked under the ladder; when the stern comes down, the wash shoves an approaching diver quite a few feet away. Add the oscillating motion of the ladder imparted by the boat's roll, and it gets really hard to anticipate where and when the ladder is going to be, and how close the diver should take a chance on approaching. The trick is to grab the ladder without getting beaned, then holding onto it long enough to get your feet—or your knees—on the rungs. Unfortunately, there's no sure-fire procedure to follow; it's strictly catch as catch can. Under certain circumstances, the best approach is to stay on the boat.

Having successfully attained the ladder is not the same as being securely seated on the boat. Oftentimes, the transfer from ladder to deck is the most precarious part of the exit. This is because once you run out of rungs there's usually nothing else to hold onto. The final pirouette over the gunwale or transom may land you on your derriere, singing a painful tune. (If you're from the British isles, it might conceivably be a London derriere.) If there's no place to install a handrail or stanchion, someone will have to be posted at the ladder in order to help people over the last step.

Some commercial operations advocate the use of a swim platform. This is a convenient place to get dressed and allows the diver to simply step off into the water in a giant stride entry. It also makes for an easy exit because the platform is nearly even with the surface: there are no high gunwales or transoms to climb over. However, I was on one boat that did not have a ladder from the platform into the sea. I had to scramble onto the steel gridwork with the aid of the biggest wave, then wallow on my belly like a beached whale until I could get my legs underneath me and push up to a kneeling position: an exercise that usually required help.

Protect yourself from harm and the elements. Most boat injuries occur below the knees: banged shins, stubbed toes, and splinters. Wear sneakers or rubber soled shoes for grip; leather soles and bare feet slip too easily, especially when the deck is wet. Open boats are mighty cold because they move so fast, creating a wind chill factor even when the air is still. Carry long-sleeved wind protection, or better yet, a rain suit.

Approaching a boat ladder can be tricky. At left, one diver grabs onto a lower rung while his buddy hovers nearby but out of the way. Middle and right: Hold the ladder firmly, then quickly get your feet or knees onto a rung, where your position is fairly secure. Climb up with your fins on.

Protect your equipment as well. Don't leave gear lying around the deck where it can trip people and get damaged in the process; pack it away immediately after each dive. Store tanks in racks if they are provided, and make sure the bungees are snug. Mark tanks and accessories with indelible ink, so you can tell them apart from identical items owned by other divers; the marks do not have to be obvious or unsightly. Plastic caps for tank valves will keep out salt spray. Mesh tank protectors not only preserve the finish on tanks, they soften the blow when tanks are accidentally banged against gunwales and railings.

You can also help to protect the boat. Tank boots prevent nicks and chips to painted surfaces and fiber glass laminate, as do vinyl coated lead weights. (Get tank boots with drain vents or a gridwork bottom so sea water can drain out and so fresh water can be flushed through when the tanks are submerged for filling.) If there is any wrath worse than a woman scorned, it is that of a captain whose boat is marred by inconsiderate divers. Lay your hardware down gently, as if you were placing it on your kitchen counter. You might get invited back.

Boats have rules. Find out what they are before you agree to dive with someone (if it's a private boat) or sign up for a charter (if it's a commercial operation). If the rules are not to your liking, look elsewhere for a way to reach your favorite wrecks. Some boats have policies so strict and diving regimes so restrictive that they are run like microcosmic dictatorships. If you are psychologically unsuited (as I am) to being treated like a serf, don't put yourself through the torture test. You're supposed to be having fun.

On the other hand, it is considered proper etiquette to inquire about the placement of equipment, dressing procedures, estimated times of arrival and departure, entry and exit points, areas that are off limits (like the engine room and wheelhouse), and the location of safety devices such as life preservers, fire extinguishers, and throw lines. It is just as necessary to know when to help as to know when to stay out of the way. The more you learn about boats, boating, and boat people, the safer you will be, the more rewarding will be your dives, and the more enjoyable will be your interaction with the wreck diving community.

Big boat diving is very much like diving off small boats, the main difference being the height of the deck, which makes for a traumatic entry and a more physically demanding exit. Although some instructors prescribe donning tanks in the water, this method is unconventional when diving wrecks due to the inordinate amount of accessory equipment that wreck divers are wont to carry; and anyway, it works only under perfect conditions in warm water with no wind, waves, current, surge, or swells.

When you jump off the boat to make your dive you should be fully dressed and ready to go underwater, with cameras, spearguns, and scooters being the only exceptions. Have your BC or drysuit partially inflated—enough to bring you back to the surface after the initial submergence, in case there is a problem, but not overly inflated because the sudden shock may stress the seams. Your fins should be secured to your feet. Your mask should be sealed to your face and held with one hand to prevent it from dislodging as you hit the water. Your regulator should be in your mouth. Hoses, gauges, and dangling accessories should be clear of obstruction.

Look before you leap. Make sure no divers or lines are underneath. If the boat is rolling, time your take-off so your flight path does not take the great circle route. Take a deep breath and hold it—in case the mouthpiece tears off from the force of entry; inhaling sea water is a sure way to go into coughing spasms. Then go for it. Unless you're doing a giant stride entry from a standing position, hit the water with your back. There's no reason to close your eyes because your mask will protect them. In fact, once you train yourself to keep them open, you'll rather enjoy the experience as you pass through the air-water interface: it will be like watching home dive movies.

After bobbing to the surface take a tentative breath from your regulator; if air flows into your lungs instead of water, breathe normally. Listen for leaks. Orient yourself with respect to the boat. Make sure your accessories are still attached. If you're planning to take pictures or spear fish, reach up and take your camera or speargun from whomever was designated to hand it to you. Next stop, the wreck.

After enough experience you can do what I call a "cat" entry. I go in negatively buoyant, and as I sink below the surface I twist like a cat in a smooth, graceful motion, ending up in a prone position and facing the anchor line, kicking. Don't pump your legs like you're riding a bicycle; that kind of kick is inefficient and doesn't make full use of your leg muscles or the design of your fins. For full power use a straight-leg kick that comes right from the hip, and strike out for the anchor line. After a while it will become second nature.

Diving from boats is not to be taken for granted. People are so used to placing absolute trust in their vehicles and expecting total satisfaction, that they sometimes forget that machines are not always one hundred percent reliable. Sometimes emergency repairs can be effected at sea, but I've been towed in on more than one occasion. I've been on boats that have suffered from a wide range of calamities such as clogged filters and fuel lines, burst cooling jackets or water intakes, engine seizure, separated drive shaft, lost propeller, drained batteries, broken transmission, broken starter, broken bilge pump, broken shaft bearings, and an untold variety of engine breakdowns. I have also been on boats that ran out of fuel and caught fire.

Think about this before investing in a boat of your own. In the long run, it may be cheaper to pay the charter fee, do your dive, and go home that night with no worries about mechanical problems and repair bills. As I said in the beginning of this chapter, boats are a necessary evil—but you do have the option of choosing the degree of evil to which you must be subjected.

Helicopter diving. Now there's an idea . . .

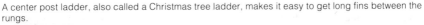
A center post ladder, also called a Christmas tree ladder, makes it easy to get long fins between the rungs.

The Art of Diving Shipwrecks

A dive is a distinct unit, with a beginning, a middle, and an end. Furthermore, it is preceded by a plan and followed by discussion. Thus, a "dive" becomes a quantum of experience. By keeping an accurate log, or diary, of each dive, you can refer to it later—years later—and jog your memory about the wreck, its contemporary condition, the depth, the temperature, what you saw and did—in short, everything you might want to remember about the experience, good and bad. You might even recall the names of your buddies and reminisce about that singular day at sea.

For log keeping purposes, the actual dive commences not as soon as you hit the water, but when you begin your descent. Thus the time you spend on the surface (adjusting equipment, swimming to the down line, or waiting for a buddy) does not count for anything. It's the same as sitting on the boat or wading through the surf.

This means that the so-called "bottom time" entered in your log does not begin when you alight on the wreck or touch the actual bottom, but includes the time it takes to get there through the water column; on the other hand, bottom time ends the moment you start up the anchor line or, in the case of an unassisted ascent, when you leave the highest point of the wreck. (On wrecks with high relief, you may never go all the way to the "bottom," but may confine your exploration to the upper decks.) This protocol is a function of decompression schedules and their inherent design, which track nitrogen onloading whenever it occurs. Inert gas is absorbed by the tissues at all times the body is under pressure, which is why descent and ascent times are critical for establishing no-decompression limits and for calculating decompression requirements.

By this reckoning a slow descent counts toward bottom time, so it behooves you to drop as fast as your ears will clear. The ongassing clock ticks throughout the entire dive, counting down to the no-decompression limit and beyond. We want to avoid the "beyond." Before you reach the maximum dive limit (the no-decompression limit, or no-D limit), beyond which you can no longer surface without stopping to offgas, you should terminate the dive and make a controlled ascent: the rate to be determined by the brand of decompression table used for the dive.

It is vitally important to maintain the proper rate of ascent, especially when you are near the maximum dive limit. Too fast, and you incur the possibility of contracting decompression sickness (DCS) due to the fast bubbling off of nitrogen. Too slow, and you add to your bottom time, which, if not properly accounted for, also may lead to DCS; this because nitrogen is still being absorbed during the ascent, even at shallow depths. The reason that decompression stops made at staged intervals near the surface are effective in preventing or at least in reducing the chance of DCS, is that the rate of offgassing exceeds the rate of ongassing, resulting in a negative net balance of nitrogen over time.

For example, the prescribed rate of ascent for the U.S. Navy Standard Air Decompression Table is sixty feet per minute. If you dive to ninety feet for thirty minutes and ascend directly to the surface at a constant speed over a period of one and a half minutes, according to the Tables you will not incur a significant risk of DCS. If you lollygag along the way, however, and take *three* and a half minutes, the two minutes in excess of the allowable ascent time must be added to your bottom time. There is no decompression schedule for thirty-two minutes. In such a case the instructions require you to round "up" to the next longer bottom time, in this case forty minutes, which necessitates a decompression stop at ten feet for seven minutes.

But we want to avoid decompression. The way we do this is by not "pushing" the limits: that is, by not waiting until the last possible minute before initiating ascent. It is important to understand that no set of decompression tables operates with pinpoint accuracy. There are so many variables involved in decompression theory that no one can predict precisely when DCS will occur. All tables (and computer simulations) are based upon a probability curve. This means that the closer you are to the maximum dive limit, the greater chance there is of getting bent.

To be on the safe side, terminate the dive well before you reach the maximum dive limit, and do a safety stop at fifteen feet for three to five minutes. These simple practices will increase your odds of avoiding DCS.

(The only foolproof method is to give up diving.) By leaving the bottom early you not only place yourself farther from the theoretical boundary of DCS, but you have some extra time in case you can't maintain the proper rate of ascent. (Your dangling accessories might become entangled with the anchor line, you might have to stop and help someone on the way down adjust his gear, or you might have to wait while a reverse sinus blockage clears itself.)

Long before you reach the maximum dive limit, however, you might have to terminate the dive due to a shortage of air. By "shortage" I do not mean that the needle of your tank pressure gauge is resting on the zero pin. The alert diver monitors his gauges continuously throughout the dive, and watches his time, depth, and remaining air supply; he also maintains a constant awareness of his position with respect to the anchor line (both distance and direction). These are data he needs to know in order to make appropriate judgment calls.

Not all wreck dives demand such strict attention to detail. When I dive in Bermuda where the wrecks lie on reefs at a depth of twenty-five feet, I tend to relax the rules. To conserve air, I snorkel until I see something interesting, then dive down on scuba and explore the area closely. If I want to move to a different part of the wreck, I surface and snorkel there. Decompression is not a factor, and a dive is not the self-contained unit that I described at the beginning of the chapter. I bob up and down with impunity. I re-orient my position every time I surface, and when my tank gets low I snorkel back to the boat.

Such is not the case for most wreck dives. Depth and limited visibility compel a more conscientious approach. Once you are well into the descent or partway through the dive, you are committed. This doesn't mean that you can't terminate the dive immediately or at any time soon after; it means that once you return to the surface, any subsequent descent that day is considered a repetitive dive for decompression purposes—unless you get back down within ten minutes. Anyway, in many cases you can't simply recontinue a curtailed dive because you no longer have a full tank of air. Your best bet is to don a fresh tank and start a new dive.

There are mitigating circumstances, of course. Once I was diving in coal-black water in which I could barely make out my elbow. My good friend Nike Seamans opted to reel out a line. Together, we felt our way along the wreck until I decided to strike out on my own. There was no way I was ever going to find her again in the blackness, but since the water was only sixty feet deep, after half an hour I ascended to the surface, swam back to the anchor line, went back down, traced out Nike's line, sneaked up behind her, and spooked her. After she overcame her fright, she began to think about how incredible my sense of direction must be. Back on the boat I bathed in her praise for quite a while before telling her the truth. She had just as much respect for my playfulness and my sense of humor.

In most situations, though, the "thirds rule" applies. Stated simply, this means exploring away from the anchor line until you've breathed one-third of

Here's clear water down to the top of the wreck, but ground swells have silted up the bottom ten feet.

your air, wending your way back on the second third, and keeping the last third in reserve. If you get back early, you can always use up some more air in the vicinity of the anchor line. If you're diving in fairly shallow water, if you do not approach your maximum dive limit, and if the current on the surface is slight, you can pretty much dive to your heart's content and come up from wherever you happen to be when you reach the final third.

If you cannot find the anchor line, don't panic. But also don't keep looking till you've seriously compromised your maximum dive limit or your air supply. Take positive action before you get yourself into trouble. Maintain neutral buoyancy, face into the current, and kick with slow, steady strokes: this will help maintain your position with respect to the boat. Closely watch your depth gauge and timer so you can monitor your ascent, and release air from your BC or drysuit accordingly. Keep an eye out for the anchor line in case you drift past it. As you near the surface, look up to make sure you don't smash into the bottom of the boat, as this could cause serious injury and possible unconsciousness.

If you are ascending with a buddy, one way to maintain contact is to shake hands while the unoccupied hand of each buddy controls his power inflator. In an air-sharing situation, hold onto your buddy's high pressure hose: one way of ensuring a good grip so you don't get so far apart vertically that the regulator gets pulled out of the mouth of the out-of-air buddy.

When you breach the surface, spin around and locate the boat, pat the top of your head to signal you're okay (if you are), assume a horizontal position, and kick your buns off. Always aim toward the bow of the boat or slightly ahead; this will enable you counteract the current while drawing near the boat on a perpendicular course. (In canoeing, this is called an "upstream ferry.")

If you see that you can't make it because the current is too swift for the distance you have to travel, keep trying; you'll probably make the trail line. If you start to get winded, slow down. If you begin to gasp—stop, lean back, get your chest out of the water, and catch your breath. Okay, so you goofed, and you're going to have to be rescued. It happens to everyone eventually. You'll have to take a good-natured ribbing, be the brunt of a lot of jokes, but you'll live through it. Just keep reminding yourself that your distress is emotional, not physical. It's an experience you'll never forget—because no one will let you.

Descending an anchor line, as simple as it sounds, requires a certain finesse, and for that reason a few words of explanation may save you some awkward moments of embarrassment and frustration. A line in the water—any line—is always a double-edged sword. Even a line that is placed there for your safety presents the possibility of entanglement because you have to touch it in order to use it. The first thing you have to understand about ropes and lines is that they are living, malevolent creatures who delight in causing grief. Man has tried to tame them, but has been only partially successful. You can't change the weave of a rope any more than you can change the stripes on a tiger; they are inherently wild and can turn on you at any moment no matter how well you treat them.

The first rope you have to tangle with—quite literally—is the granny line. It floats in the water sinuously, almost sybaritically, swaying to the rhythmic beat of the sea. You reach out for it, take hold of it, pull yourself hand over hand, and—it grabs you by the leg and wraps around your hoses and dangling accessories. Suddenly you're trussed up like a fly in a web. Intellectually you know that an errant wave flipped the rope over top of you, then the current pulled it taut. But deep down inside you know the rope was waiting for you, stalking you: for divers are its prey.

Anchor lines are just as predatory. In the same manner in which pitcher plants attract insects into the fold, anchor lines hook themselves into wrecks in order to prey on unwary wreck divers. If you stay on either side of the line, it will grab your spare hoses; if the line is beneath you, it will twist your goody bag in knots; if you try to get under the line, it will grab your tank valves. Once, the anchor line insinuated itself between the valves of my main tank and my pony bottle. I struggled in vain to free myself, pushing and pulling frantically before the spider appeared to deliver the paralyzing sting. Finally, I went limp and let my buddy extricate me.

Nets and monofilament are even more insidious. They lurk on the wrecks in unsuspecting stillness, leading you to believe that they are asleep and that you can quietly sneak past them without being noticed. Just when you think

Nets and monofilament are perennial wreck diving hazards.

you're in the clear, they lash out with an invisible tractor beam more difficult to escape than the one on the starship *Enterprise:* it's called surge, and unless you have a surge protector you're bound to get knotted up worse than a dropped cat's cradle. Be specially careful of nets that billow in the current, for they are notorious for snaring divers.

So where did all these nets and the monofilament come from? A ocean wreck is an artificial reef; the wood or steel provides substrate for marine organisms which start a food chain that works up to the largest predators and passing pelagic species. Sportfishing on wrecks is a popular pastime, and commercial fisherfolk regularly frequent the wrecks in order to earn a livelihood and to provide delicacies for landbound restaurants and super-markets. In human terms, the top end of this chain, the last link before the lock is snapped in place, is called consumerism.

Sometimes the quest for fun and food goes awry. Hooks snag in the wreckage and the lines snap. Nets hang up and can't be retrieved; they break or have to be cut free. Fish traps and lobster pots are laid near wrecks because of the local concentration of game; if a storm blows them too close to the wreckage, the connecting lines become fouled and the traps and pots cannot be retrieved. But, these nets and traps and pots continue to do the job for which they were manufactured: they catch game. The trapped animals die, their rotting flesh entices other animals that soon become enmeshed, and the deadly cycle continues without any perceivable end.

Then divers come along to spear a few fish and grab a couple of lobsters, maybe collect some scallops or abalone. But the amount of biomass removed by such handwork methods is inconsequential when compared to the amount taken by commercial enterprise. In fact, much more biomass is *wasted* by commercial accidents such as those described above than divers could ever possibly reap from the wrecks.

You can practice good shipwreck ecology by taking a few minutes of your dive to help rectify the unfortunate results of fishing gear losses. If you see a fish caught in the tatters of a net, cut it loose, then hack the net apart to prevent further bloodshed. If you see lost fish traps and lobster pots swarming with game, open the doors and let them out; bash in the slats of wooden traps, lash open the doors of wire traps. You may keep the fish and lobsters, of

Untethered traps should be smashed open in order to end the senseless killing cycle.

course—consider it your reward for being a Good Samaritan. (On the other hand, don't bother a trap that is off the wreck and free of entanglement, and don't steal from it. A properly set trap is private property and should be left alone.) You might also want to slash those draping strands of monofilament—but that's more for the divers' benefit than the fish's.

Perhaps the two most important skills to master in wreck diving are recognition and navigation. Both are interrelated and interdependent: recognition is an aid to navigation, and vice versa. Recognition can be separated into two major components: a basic understanding of ship construction and identification of parts, and the recognition of these features when they are segmented or thickly covered by marine encrustation. The first comes from studying photographs, drawings, models, and ships themselves (many major ports have ships as tourist attractions). The second comes only from experience in the water.

For example, the ends of a wooden sailing ship are distinctive: the bow is pointed, the stern is rounded. But after a hundred years of deterioration, collapse, and partial burial, these features may no longer be well defined. Thus if you recognize a pile of heavy iron links as anchor chain, chances are you're in the bow; if you determine that a large section of timber is the rudder, you're in the stern. Furthermore, these identifiable features become landmarks from which you can orient yourself with respect to the anchor line and to the rest of the wreck.

Another simple rule, this one for orientation on steamships, is that the boilers are always forward of the engine. If you come upon the steam engine, look around for the boilers; they point the way to the bow. By corollary, if you know that you're headed for the bow you'll be better prepared to recognize that half-buried, bulbous bronze object as the ship's bell.

It is helpful if you know where you are on the wreck at all times during the dive. It puts you "in the picture," so to speak. Instead of just wandering aimlessly on a mass of metal plates or a pile of wooden beams, you develop an image of the wreck as a whole. By seeing how the various parts are placed with respect to each other, you'll be able to visualize the wreck as the remnants of a ship, much as you would behold a completed toy puzzle as a recognizable scene instead of as a bunch of jumbled carved pieces. With this

Dead giveaways: anchor chain denotes the bow, the propeller the stern.

The anchor line is hooked into the engine—the tall mass at the left. Lying on its end, the cylindrical boiler points the way to the bow.

growing appreciation of a shipwreck's likeness comes the same sense of fulfillment that comes from solving a complicated riddle or piecing together the parts of a table-top puzzle. Thus wreck diving is a stimulating, intellectual challenge that tests your skills of observation and powers of retention.

Diving in wonderful tropical clarity gives you an edge in forming an overview of a shipwreck's appearance. You may be able to put the scattered pieces into perspective on a single dive. But in darker water, where particulate matter obscures omniscient observation, you may have to dive the wreck many times before you gain the same understanding of its layout. One river wreck I dive has visibility ranging between one and three feet. Looking in all directions from any spot on the wreck, I could see at best a six-foot swath. Consequently, it took quite a number of dives for me to piece together the individual swaths into a comprehensible whole—and even then there were gaps in the image because I could never remember all the swaths and how they fit into the pattern. My mental image of the wreck was always out of focus and full of conjecture.

To formulate a picture of a wreck in my mind, and to jog my memory about what I saw on the dive, I make sketches. Putting my initial impressions on paper helps me to remember peripheral details that might otherwise be forgotten in the flood of raw sensory data: it forces my subconscious to divulge background particulars. On day trips I might make my drawings after I get home, especially if it's been a rough day at sea and the exigencies of getting back to port in one piece have kept me otherwise occupied. However, I always have a tablet with me, and if possible I materialize my images with ink or graphite while they are still fresh in my mind. For better clarity, I might redraft the sketches later when the wood under my feet isn't moving.

If I'm conducting an actual survey I get a bit more sophisticated. I take a slate and grease pencil into the water, and draw the wreck as I explore it. The act of depicting the wreck the way it appears from various perspectives compels me to notice more than I would with a cursory glance at the structure while looking for hidden lobsters or portholes. I don't profess to be an artist, but art is not my purpose. Once I know how a wreck is laid out, the next time it will easier to find my way around. I flesh out the survey by snapping pictures of objects that are too complicated to draw; I indicate on the slate where the photos were taken and at what angle and elevation, then add the details to my master drawing after the film is processed. And if you really want to be precise, take along a tape measure.

If this approach to wreck diving seems too systematic, it is necessary to ask yourself why you are diving wrecks instead of quarries or coral reefs. If your true purpose is to explore these sunken windows into the past, and to glean something from the experience other than to soak some very expensive equipment, then the systematic approach will prove the more rewarding. This is not to demean less demanding objectives such as hunting and collecting shells, it just puts the various goals into perspective.

Now let's talk about the other important skill necessary for wreck exploration no matter what your purpose is in being there: navigation. I don't mean piloting ships across storm-swept seas by taking fixes off the stars, but what on land in wilderness exploration is called "orienteering": that is, starting out from a familiar spot, checking out the neighborhood, then returning to the spot from which you began. People who have only dived in the Caribbean will undoubtedly scoff at such a notion, because it's difficult to get disoriented where the water is as clear as air. But the loss of visibility in the northern oceans and lakes makes accurate navigation an essential proficiency to have.

Not only will knowing where you are all the time enable you to relocate the anchor line, it will reduce the amount of stress that results from being in an otherwise unpredictable environment: the less you have to worry about, the more you can concentrate on other matters. (Remember how you felt the last time you got lost in big city traffic or on a lonely country road? It can be unnerving.) Don't be afraid to admit that you've been lost or disoriented on a wreck. It happens to everyone. Daniel Boone, one of the greatest scouts and woodsmen of all time, was once asked if he'd ever been lost in the forest. His apocryphal quip was, "No, but I been a might confused fer a few days."

Shipwreck navigation requires three basic items: a compass, a wreck reel, and common sense. The most important is common sense. Most of the time you can dispense with the other two items because they are not warranted by the conditions. With decent visibility you need only a good sense of direction and an eye for landmarks; you can rely on natural means of navigation.

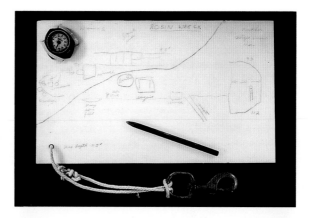

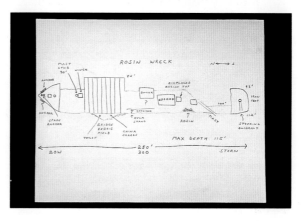

My underwater sketch on the slate had to be compressed because of the length of the wreck. That night, with the details still fresh in my memory, I redrafted the rough sketch to produce the one on paper.

If the light penetration is strong enough you can use the angle of sunbeams carving through the water as an aid to direction, or the sun itself. If the bottom is sandy you can use the ripples in the sand for exploring short distances off the wreck; for example, if you cross the ripples on the perpendicular, it's necessary only to turn around and swim back on the reverse perpendicular, and you will end up very near your starting point.

Current is also a good guide, for it seldom changes during the course of a dive. If you swim away from the anchor line with the current on your right, return by swimming back with the current on your left. If the current is very strong, however, it is always best to swim *into* it at the beginning of the dive so it will be at your back when you turn around. If you swim with the current at first until you breathe one-third of your air, you may need considerably more than your remaining air to make it back to the anchor line, if you can make it at all. Furthermore, if you don't relocate the anchor line and have to ascend without it, you'll come up adjacent to or behind the boat instead of in front of it. By swimming into the current from the start, an unassisted ascent will at least place you far enough in front of the boat so you can vector back to it with the drift.

On shore dives you begin at the surface level and work your way into deeper water. If you get completely disoriented, you can usually follow the slope into the shallows. However, along rocky coastlines there may be drop-offs instead of a slope, in which case you'll have to follow sunbeams or ripples, or utilize a compass. You can also take advantage of the surge.

Shallow-water surge is caused by the passage of waves along the surface. As wave heights rise and fall, some of the energy of movement is transferred to the horizontal plane, creating a back-and-forth motion that can be quite extreme. I've experienced surge in which the water mass moved as much as ten feet in one direction, stopped, then moved ten feet back in the opposite direction. The fish don't seem to mind; they simply go with the flow. But for divers trying to maintain position with respect to the wreck, the constant state of flux can be disturbing, even dangerous. Once you've been slammed against the side of a rock or a steel plate, you'll know what I mean. Extreme caution must be exercised when approaching sharpened metal beams or protruding timbers; not only can they impale you, they can snatch your hoses or dangling accessories as you speed by, and do serious damage or whip you against the wreck. You can also get pinned against a wreck or rock, stuck like a moth on the front grill of a car.

I've also experienced deep-water surge at depths of over a hundred feet. The side-to-side motion is not as pronounced nor is the force of moving water as strong, but the presence of a shipwreck can produce some startling effects. If the hull is fairly intact but riddled with rust holes, the surge forces the water through the body of the wreck. During one cycle, the water pours into the holes on one side of the wreck and out the holes on the other side; during the reverse cycle the water flows back the other way. Generally, this kind of wreck is like an iron Swiss cheese, with holes that are not uniform in size. Thus a massive flow of water may enter the hull broadside through an incredibly large opening, then be forced through a bunch of little holes on the other side. In either direction this surge of movement creates a siphon effect. To a lesser degree, this can also occur on isolated hull plates.

What's it all mean? I'll cite a couple of examples from my vast font of experience. Once while passing a glassless porthole along the side of a wreck, my arm was sucked into the opening and my face was plastered up against the barnacle encrusted exterior, dislodging my mask and causing it to flood, and twisting my regulator in my mouth so that I could not inhale without taking in water. I was stuck there for five or ten seconds, until the surge reversed and released me—then I was shot out like a watermelon seed in a spitting contest. Another time I was passing a doorway in a washout when the surge grabbed me, spun me around three times like a doll in a cyclone, and deposited me on the bottom some fifteen feet away; I felt as if I'd just stepped out of a commercial dryer.

On the other hand, instead of always being victimized by the surge, you may be able to make it work for you. Since the direction of reversal is constant, for navigational purposes you can treat surge the same way you would treat current: with a constant mental awareness of its orientation to you and to the wreck. Surge on near-shore wrecks can guide you back to the beach or rocks because the waves that generate it seldom move parallel to the coast; waves may not always break perpendicular to the coast, but the angle of incidence will at least be in the proper hemisphere. Diving in the kelp beds off the state of California, a valley girl would undoubtedly say that the surge aims "for shore, for sure."

Another way of maintaining your orientation is by swimming in straight lines. If you meander too much you may not go around in circles, but you'll create so many curves that it will be difficult to retrace your path. On large, broken-down wrecks, swim along the edge until you reach a point where you want to go inboard, enter the inner debris field at a right angle to your original line of travel, explore the area, then return to the edge of the wreck which you can follow back the way you came.

You can also leave a marker on the edge of the wreck when you begin your incursion. This can be either a strobe or a light, the idea being to keep it in sight at all times. If you get out of range you may not only lose your way, you may lose your expensive marker as well.

You will soon learn that the wreck can look entirely different when seen from different directions. If you're unfamiliar with the site, it pays to look over your shoulder every once in a while in order to see what the landmarks look like from the other side.

If you dive a wreck consistently you will eventually get to know it. This is an advantage because after a while you'll be able to navigate on it just as easily as you can find your way through your house in the dark, or drive in your neighborhood without looking at the street signs. Not only will you feel more at ease on the dive, but it will deepen your appreciation of the wreck while enabling you to discover its innermost secrets: like where to dig for artifacts, or where that hole is that always has a lobster in it.

If you plan to "work" a wreck (that is, visit the site on a periodic basis), you should consider establishing a permanent marker tied to the area you want to explore; that way, each dive will begin from a familiar spot and you will be pre-oriented upon reaching the bottom. (It is this fact that makes more efficient a repetitive dive on the same wreck, rather than moving to a second wreck and

This large boiler makes a recognizable land-mark.

grappling blindly.) If the visibility is commonly poor, you can place guidelines across the wreck in order to take the guesswork out of getting from one place to another.

I have dived some wrecks over a hundred times and have never tired of them. Like growing children, they change throughout the years and maintain a continuing fascination.

Another way of honing your shipwreck orienteering skills is by practicing navigation by compass. If you learned how to use a compass in the scouts, you're way ahead of the game, for it's no different underwater. Total submergence does not shield the needle from magnetic lines of flux. The place to begin using a compass is on shore before you wade into the breakers, or on the boat before the dive; this is to establish your position with relation to the most important known reference point: the place you want to return to.

If the local coastline trends northeast-southwest, you begin the dive by heading southeast (offshore) and return by swimming northwest. The bezel will help you remember. On land, steady the compass till the needle points north, turn the compass bowl till the index mark also points north, then rotate the bezel so the lubber line points to 135° (southeast). After you enter the water, all you have to do is realign the needle with the index mark, and follow the preset lubber line. It's not as easy as it sounds, and will take a bit of practice before you get the hang of it. Make some trial runs in your back yard before you try to do it on a wreck.

To get back to your starting point you "shoot a back azimuth"; that is, the reciprocal bearing that is 180° from your original course. In the example above, you add 180° to 135° and get 315° (northwest). If your initial bearing is more than 180°, you get the reciprocal by subtracting 180°. Compasses designed for underwater use generally feature a reciprocal mark so you don't have to do any arithmetic in your head.

For boat diving, wait till the boat swings around after the grapnel has been set, then take a bearing on the bow (or, more accurately, off the anchor line—the boat may be stern anchored). Since the anchor line points toward the wreck, you'll know when you reach the bottom which way to go in order to stay in front of the boat. Keep this bearing in mind or write it on your slate.

The compass is particularly useful on wrecks that have little definition and no outstanding features, or that consist mostly of widely scattered debris fields separated by pockets of sand. Choose a direction from the grapnel, take the heading from the compass, sight an object that is in that direction, and go for it. Keep repeating the operation until you decide to turn around and run the reciprocal course.

Many ships have hulls made of iron or steel, and many wooden hulled ships have iron steam engines, auxiliary machinery, or other metallic parts. The wrecks of these ships disturb the local magnetic field the same as a natural concentration of iron ore, and affect any compass in the vicinity. This local disturbance is called "deviation." The closer a compass is brought to the cause of the disturbance, the more will be the deviation. This means that the compass

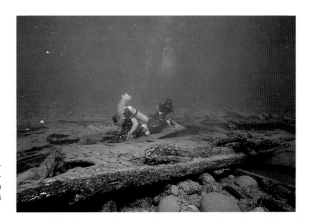

On widely scattered wreckage a compass can be useful; if this wreck existed in dark, dingy water a compass would be essential.

needles of passing ships and boats will be deflected by an amount determined by the strength or intensity of the disturbance. These localized deviations are trivial with regard to marine navigation, but can affect a compass that is practically in touch with the wreck.

By this exposition I do not mean to imply that a compass will not function on an iron or steel wreck, only that it will be less accurate. Magnetism is a weak force whose interaction diminishes greatly with distance. As an experiment, place a dense iron or steel mass (pliers, scissors, or hair spray can) next to the bowl of a compass. The needle instantly swings around and locks itself on the external mass. As you move the object away, the needle swings slowly back toward magnetic north. At a distance of one inch the deviation is only a few degrees; at two inches the deflection of the needle is barely noticeable; at three inches there is no affect.

For our purposes, too, it's important to understand that as a wreck degrades, it loses its inherent magnetism. Let me explain why. Magnetism is produced by the alignment of iron atoms which are themselves magnetic: that is, each atom of iron is in effect a tiny magnet. If iron atoms were arranged haphazardly, or dumped on the ground so their alignments were determined statistically by the way they landed, the atoms would point in an infinite number of directions (in all three dimensions) and cancel any overall magnetic effect: the net charge would be zero.

The Earth is a giant magnet, presumably (according to modern scientific theory) because it has a molten iron core whose eddies create an electromagnetic field. Invisible lines of force span between the north magnetic pole and the south magnetic pole. The Earth's magnetic field exerts a force which causes iron molecules to align themselves with the planet's magnetic poles, thus prevailing over the statistical arrangement described in the previous paragraph.

You can turn a screwdriver into a weak magnet by holding it north/south and banging it with a hammer. As the molecules "jump" within the solid matrix, they have a tendency to "land" in alignment with the Earth's magnetic poles. This is called "induced magnetism." (Conversely, you can destroy the magnetic properties of a magnet by holding it east/west and banging it with a hammer.) Similarly, ships built on a north/south launching ways will generate a stronger magnetic field than ships build on an east/west ways. (Must be all that banging that goes on during construction.)

Finally, the last ingredient that goes into this recipe is iron oxide, or rust. The addition of oxygen atoms to the iron matrix destroys the magnetism on the atomic level (called the "dipole moment") and consequently diminishes the cumulative magnetic effect.

Result: the older and the rustier the wreck, the "softer" is its magnetic signature, and for a variety of reasons. The iron atoms will unalign themselves over time, due to the disturbance or "banging" effects of earthquakes, wave action, depth charges, and so on. As hull plates fall off and machinery is displaced, the resultant disarray reduces the strength of the wreck's original magnetic field. And, as the iron components oxidize, magnetism degenerates at the atomic level. Given enough time, a shipwreck will reach equilibrium with its surroundings and become nonexistent.

In the short version, this means that older wrecks will not affect a compass needle as much as recent wrecks. This is to your advantage. Old wrecks exist in a state of collapse and will require proficiency in the use of a compass under shipwreck conditions in which a compass is the most effective, whereas recent wrecks are likely to be intact, enabling you to emphasize recognition methods for navigation.

Notwithstanding this degree of difference between old wrecks and new, for all practical purposes a compass works remarkably well under most shipwreck orientation circumstances. For one thing, although the local magnetic field will cause deviation of the compass needle, that deviation will remain fairly constant over the entire site. Even if the compass needle completely reversed itself in the presence of a wreck, within that framework you could still use the compass to get from one point to another. Furthermore, as demonstrated by the experiment in which a mass of metal deflected a compass needle, as long as you don't put the compass bowl in contact with the wreck, the needle will not be noticeably affected. In short, for navigational purposes you can use a compass with confidence.

When you use a compass, hover a few feet above the wreck and keep it as far away from your equipment as possible. This means that you can forget about watch-band compasses whose lettering is too small to decipher without having it right in front of your mask. Get a compass that you can read at arm's length. Also, get one whose masks are dabbed with luminescent paint that is visible in low light, or that can be "powered" by the flash of your light.

A compass needle can be deflected by the proximity of electricity. This is because magnetism and electricity are related phenomena. Thus, electricity can induce a magnetic field, and magnetism can induce the flow of electrons. If you strap a compass on your wrist, keep your light away from it when you are taking bearings. By the same token, a compass mounted on a console will not work at all if a battery-operated timer or decompression computer is also mounted on the console; in fact, the compass needle may point only at the electrical source and nowhere else.

However you carry your compass, make sure to keep it perfectly level when sighting from it. If the needle touches the lens, the friction will prevent the needle from moving.

In extremely turbid water a compass will not help you relocate the anchor line, no matter how good you are at orienteering. I was once in water so black and filled with sediment that it took several minutes of groping to find an anchor line that was touching my leg. Under such conditions, there's only one way to ensure a direct return to the boat: physical contact.

Obviously, you can't drag the anchor line around when there's a boat attached to the other end of it. But you can extend the continuous guideline to the surface by attaching to the grapnel a line of your own. This is called a "way-back" line or a "cross-wreck" line; it is carried on a plastic or aluminum line reel whose spool can hold several hundred feet of eighth-inch nylon, and which can be rewound the same way you pull in a flounder with a fishing reel.

The line reel comes with an angle bracket for a handle and a snap hook for attachment to a D-ring. You can put another snap hook on the end of the line so you can either clip it to a link in the anchor chain or wrap it around the grapnel shank and hook it onto the cross-wreck line. The advantage to this arrangement is that, should the grapnel pull free from the wreck, you'll be the first to know about it. However, you must be able to let go of the reel lest you get dragged through the water like the catch of the day; this means no lanyard on your wrist. Depending upon conditions and the security of the grapnel, you might opt to tie off to a piece of wreckage within reach of the anchor line.

There's no "drag" on a cross-wreck reel as there is on a fishing reel. After releasing the locking screw you have to hold your thumb against the side

The divers below have tied off a cross-wreck line high on the anchor line, and are running it to where they are working on a project. This will enable them to return straight to the anchor line, and will enable others working on the project to go directly to the work area. The diver at right carries his reel on his belt.

of the spool to prevent the line from reeling off too fast and creating a monstrous tangle. If a tangle does occur, instead of reeling in the line as you're supposed to, wrap it sideways around the handle and spool. In either case, no matter how far or in which direction you travel, you can always find your way back to the anchor line. You might say that it's "reel" simple.

In practice, it's a little harder than it sounds, but not much. If you've read the previous chapter, then you know that any time you fool around with ropes and lines you face the possibility of entanglement. It's no different with a cross-wreck line: it can snag on wreckage, wrap around your fins and hoses, and tie your buddy in knots. (No problem that a sharp knife can't solve.)

When laying line, be careful not to let it touch sharp metal edges or fall into grooves which may make retrieval difficult. Place it where it will be out of the way of other divers, who may not be reluctant to part ways when they find themselves knotted up like a doughnut box.

It's a good idea to do "secondary tie-offs": instead of having a hundred feet of line sweeping in a great arc with the current, keep the line taut, lay a loop over a convenient projection, then go on to another projection and lay another loop. It takes only a moment, and it will prevent the current or surge from entangling the line behind you.

A line reel can also be used to make search patterns off the edge of a wreck, in order to look for disarticulated fragments and associated debris, or to try to find that expensive camera you dropped on the way down the anchor line. Tie the line where you want to begin your search, swim half the distance of the visibility, lock the reel, and sweep an arc out across the bottom until you encounter the wreck again. Let out more line and sweep your way back. In this way you can cover a great amount of area without fear of losing your direction on the flat, featureless plane. You can also use the line reel to establish permanent guidelines, as mentioned above.

Line reels have their place in dark water, too. But that's the next chapter.

Dark Adventure

As the sun settled behind the distant blue horizon, the sky scintillated like amethyst—narrow beams reflecting off dust in the atmosphere—then gradually darkened from the lighter shades of cerulean to a deep vivid purple, and eventually to the color of India ink: black, swirling, and fathomless. The stars were incredibly sharp pinpoints of white pasted on the grand backdrop of the firmament.

Beneath my fins the water was invisible; there was blackness there but I could not see it. The sea was as still as rock, the current moved with the speed of granite. I felt no sensation of temperature. And gravity may have been only a theoretical concept.

I floated in motionless nullity like a disembodied mind, like an intangible perception, like a nonexistant mote suspended in the infinite vastness of space. My imagination wandered freely through the cosmos. This was no dream. This was an altered state of consciousness induced by near total sensory deprivation. My only touch with reality was my hand on a slender silver thread that purported to be an anchor line. No, this was no dream. This was a night dive.

Not that all dives after dark are so evocative. Diving at night can be an unforgettable experience, or it can be like any other dive only darker. As with all encounters in life, what you get out of an experience depends upon what you *want* out of the experience. Nor do I mean to suggest that by sticking your face into pitch black water you'll be able to trip through space and time like an acid freak on LSD. Night diving may not transport you into another dimension, but it will admit you to worlds that you never knew existed.

Nocturnal creatures seldom seen during the day will be out in profusion at night. Lobsters that hide deep in their holes come out after dark to feed. This can be either a photographer's dream or an epicurean delight: quick shooting or easy pickings, depending upon your tastes. What's the cost of this midnight rendezvous? Let me shed some light on the subject.

Man lives by his eyes, so it goes without saying that in order to see anything on the wrecks at night the darkness must be dispelled by artificial illumination. What we think of at home as a flashlight, under water becomes a "dive" light. But don't think of a dive light as the kind of flashlight you keep in your kitchen cabinet: the kind that dims and flickers with frustrating irregularity, the kind that you have to bang with your hand in order to make it work. Dive lights are extremely reliable. To be sure, batteries can go dead or a filament can burn out and occasionally a casing will flood, but with proper care and maintenance a dive light will provide years of underwater enlightenment.

There are many kinds of dive lights: from small to large, from dim to bright, from wide beam to narrow, with rechargeable batteries or throwaways,

cheap and expensive. Which one you choose depends upon the intended application.

Most divers start out with an inexpensive light powered by disposable alkaline batteries. For everyday wreck diving, when you only expect to use a light to peer under a hull plate or look into a slightly darkened crevice, or read your gauges under low-light conditions, a fist-sized model that operates on two to four C-cells is probably all you need. This is called a "mini" light. There are many kinds and configurations of mini lights; their advantage is that they are small and easy to carry. Some have negated one of the topside flashlight's greatest weak points by doing away with the switch; the light is turned on by twisting the lens cap, which action presses the contact point on the bottom of the bulb against the tip of the battery.

Even dive lights with mechanical switches are more dependable than their topside counterparts. The only disadvantage is longterm and may never manifest itself; it is due to the hole in the body of the casing for the actuator arm. Every entry point into an underwater housing requires a waterproof seat and o-ring, and increases the chance of leakage due to cracks that develop in the rubber over time, or due to extrusion of the o-ring from its seat by the pressure. If you notice minute droplets of water or rust stains in the vicinity of the inside of the switch mechanism, disassemble the switch and replace the o-ring, or send it back to the factory for overhaul.

Better quality mini lights may feature lithium cells, rechargeable ni-cads (nickel-cadmium), and may have high-intensity xenon or krypton bulls instead of the usual quartz-halogen. All of these make good lights for back-up and for occasional use. If you get one with a holster, you can mount it some place out of the way (even on top of your head) where it will always be available when you need it. If you outgrow your mini light it becomes a dandy backup or secondary.

So called mid-range lights are bigger versions of the mini lights with more or larger batteries and a reflector larger in diameter. With more power to draw on, these lights are brighter and will emit light longer before the batteries need to be replaced or recharged. The casings of these lights are still small enough to enable you to hold them in your hand.

Then come the truly large lights—those that are bigger than your grasp and which therefore require a handle to hold on to. There are two kinds of handles. The pistol grip allows you to point the light like a gun, while the lantern grip lets the light point forward from a lower angle. Which grip you choose—or find more comfortable—depends upon the type of diving you do and how you swim along a wreck.

For example, if you like to skirt the edge of a wreck and look into holes and under hull plates for lobsters, the lantern grip will most likely suit your needs better. On the other hand (so to speak), if you generally explore a wreck by swimming above it, with your body angled at about forty-five degrees so that you can look down on the wreck while seeing what is ahead, the pistol grip will be your handle of choice. The goal is to keep your hand aligned with your forearm so your wrist is not bent up or down. The proper grip will reduce wrist fatigue.

If you do all kinds of diving and are in a quandary as to which type of handle will work best for you, you can buy two lights, each with a different handle, then use the one that is the most appropriate for each particular dive. Or you can purchase a light from a manufacturer that offers interchangeable handles.

Rechargeable batteries cost more initially, but may save you money in the long run because they don't have to be replaced every couple of dives; the batteries may last for years. This can be convenient because you don't have to continually buy new batteries. You simply plug in the light after each day of diving, and let the batteries recharge. Keep in mind, though, that the trickle charger that comes with some lights may require as much as fifteen hours to get the battery up to full power. For multi-day diving, where the days are long and the nights are short, there may not be enough time to get a full charge for the next day's diving. A boat that you use for an overnight trip may not have electricity available for charging lights. And if you dive in other countries, they may use electric current at a higher voltage, or generate electricity at a lower frequency; some chargers come with a converter to take these conditions into account.

Lights powered by rechargeable batteries are significantly brighter than those that rely on disposable batteries. Furthermore, rechargeable batteries operate on full power until practically all their energy is drained. Unfortunately, this means that there is little or no warning that a light is about to fail; it dies suddenly.

Rechargeable batteries also need to be recharged every month or so even if you don't use the light; otherwise, the slow, natural drain of power may damage the battery's ability to recharge, and may reduce its life expectancy. A

Northern bugs come out at night.

diving regimen of annual trips to the Caribbean may make this kind of light more trouble than it's worth and more costly in the long run.

In short, there are many circumstances in which the advantages of a brilliant, rechargeable light are outweighed by the dissimilar qualities of a light that shines longer but with lesser intensity. You'll have less of an initial outlay. You can pack a lot of extra D-cells and change them whenever you notice a gradual dimming; you'll still have enough light to get through a dive of normal length. And you don't have to string extension cords all over your house or hotel room.

Some light casings can be purchased with both disposable or rechargeable batteries, and the required bulb for each, allowing you to switch from one system to another depending upon the circumstances of the dive or the trip on which you intend to use the light. This requires less of an investment than that needed to own two separate lights.

No matter what size or type of light you purchase, make sure to get a wrist lanyard for it.

If you take care of your dive light it will last for many years. Treat it delicately; don't toss it around like a wrench or a weight belt. Excessive jarring can loosen the electrical connections, break solder joints, crack the reflector, or even damage the casing. If it reaches the point where it works only intermittently, return it to the manufacturer for overhaul. It makes no sense to take a troublesome light on a dive when the failure of the light means terminating the dive: the cost of repair, possibly even replacement, may be less than the cost of the dive.

There are preventive maintenance measures you can take yourself that will help ensure that your light is in good working order and ready to dive at a moment's notice. Rinse thoroughly after an ocean dive in order to eliminate the build-up of salt. Not only can salt increase the corrosion rate of exposed metallic parts, but crystal growths that accumulate under switch levers and rotary lenses can retard their movement or jam their operation completely. If salt crystallizes on an o-ring, the rubber surface might be scored and the watertight integrity will be compromised.

Periodically disassemble your light, remove all consumer-accessible o-rings, and clean each one thoroughly with fresh water. If you find sand or grit or a build-up of dried o-ring grease, wash it off gently so as not to damage the rubber. Dry the o-ring with a lint-free cloth (a T-shirt tail works fine) and inspect it for cracks.

While you have the light apart, check inside for signs of water or condensation. Soak up damp spots with a cloth or paper towel. Replace or recharge batteries as needed. If I have alkaline batteries that may be weak but not fully drained, I take them out and use them in my household flashlights where their sudden failure is not a catastrophic loss. Read the instructions for rechargeable batteries. Some manufacturers suggest that they be drained completely before being recharged, so they won't take a "memory" and accept only a partial charge.

Also, rub a pencil eraser over the base of the bulb and over all battery contact points; this will wipe off the black, sooty deposition (carbon deposits) that block the flow of electrons. If the eraser is not abrasive enough to do the job, use a fingernail emery board. Carbon deposits are what make a flashlight with weak batteries flicker and go out; when you bang the light against your hand, you either scrape off some of the deposits or you realign the contact points so they touch at a clean spot.

Dab on a tiny bit of o-ring grease and spread it smoothly on the o-ring with your thumb and forefinger. Don't use too much grease: its purpose is to lubricate the rubber, not to make a watertight seal. An overabundance of grease can alter the fit between the o-ring and the seat, and create a micro-scopic "bulge" that might extrude under pressure and cause a leak; further-more, since particulate matter sticks to the grease, a thick layer of grease allows more of it to accumulate. Read the instructions that come with your light; some manufacturers use dry o-rings that need no lubrication.

Replace the o-ring carefully in the groove and reassemble the light. If the parts are made of clear plastic, inspect the o-ring and make sure it is free of

So do southern bugs.

kinks and bulges. Turn on the light to see if it works. If not, take the light apart again and recheck the connections. If the light has ni-cad batteries, do not turn on the light immediately after recharging them; at that point they are supercharged, and the sudden surge of power may blow out the bulb.

If you find small droplets of water inside that cannot be accounted for by condensation, pay particular attention when you check and clean the o-rings. Stretch the o-rings slightly and see if cracks or thin spots appear. Take special care not to pinch the o-rings during reassembly. If the light leaks again, it probably needs service from the manufacturer; some of the o-rings, like those around the switch or the external charging port, are not consumer accessible. The light has to be dismantled by a technician at the factory.

If you see water trickling into your light on the way down, return to the surface immediately and get it out of the water. Open it and drain it, then leave it out to dry. Do *not* tighten the lens or ports under water even if you think they were not completely closed; tightening at depth makes it impossible to reopen at the surface.

All is not lost if a light floods during a dive. I've even seen lights continue to stay lit despite total flooding. For the well-being of the electrical components, however, it is advisable to turn off a light as soon as you notice that it's flooded. Continue on the dive with your back-up light—you won't save anything by rushing to the surface once flooding is complete. If you have a goody bag available put the flooded light inside. This is so you won't lose any parts should the light explode during ascent.

I don't mean like a bomb or a hand grenade, although the concussion can startle you if the light lets go under water. There won't be any shrapnel. But remember that a battery is a container of charged chemicals storing potential energy in the form of electricity. When sea water penetrates the battery it reacts with the chemicals inside, forming a gas that is released into the casing of the light. An "explosion" is defined as a rapid expansion of gases; in this case we get a slow expansion that is confined by the pressure-proof resistance of the casing. On the bottom, the external pressure of the water is greater than the internal pressure exerted by the expanding gas. But during ascent the external pressure decreases and the differential is lowered. When the point is reached where the internal pressure exceeds the external pressure, the light is in danger of "exploding": the casing may crack, the threaded screw-ring may split, or the clips holding down the lens retainer may pop off. What you hear is a sharp "crack" as the weakest part "gives." This usually occurs near the surface where the ambient pressure reduces drastically, or just as you climb out of the water.

The casing may do little more than effervesce as gas and muddy water pour out of the fracture zones, or there may be significant structural damage as molded plastic assemblies split completely apart. Obviously, I'm describing an extreme case, but it does happen once in a while. If your light floods totally it's not a bad idea to prevent the potential destruction of the components by opening it on the bottom in order to reduce the stress of the ascent. After all, it

A "head" light leaves your hands free for grabbing.

can't get any wetter.

Be extremely careful when opening a flooded light on the surface. Just because the internal pressure was not strong enough to split the casing open does not mean that when you unscrew the lens holder or unclip the front port it won't fly apart in your face. Immerse a totally flooded light in a bucket of water before opening it; that will dampen the effect of the sudden expansion.

Soak the entire light in fresh water as soon as it's feasible—a couple of minutes either way isn't going to make or break it. After a thorough rinsing, shake off and towel off as much water as possible, then lay out the parts to dry. The subassemblies that are damaged the most by water (especially sea water) are the battery pack and the wiring harness. Lights without sealed beams may lose the reflective coating on the reflector, but the reflector is easily and cheaply replaced. Everything else is made of plastic. If you dry the parts before corrosion sets in, you may salvage the wiring harness and associated screws and solder joints, although sometimes they are baked when the shorted circuit shunts electricity through them. If you're lucky, all you have to buy are new batteries—which in the case of rechargeable lights are the most expensive items.

When it comes to dive light technique, there's very little to know other than not to shine a light into your buddy's eyes: not only will it blind him momentarily and wipe out his night vision, but he may retaliate in kind, or worse. It is so natural for us to communicate with the eyes, in order to read the expression of the soul, that shining a light there comes natural. Concentrate on aiming your light at a person's chest or to either side of his face. If he is looking the other way, flash your beam quickly in front of him if you want to attract his attention.

Getting back to the beach, boat, or anchor line assumes vital importance at night because it is so difficult to be seen should you lose control of the dive and drift out of range of safety. Strobe lights become invaluable in this regard. Divers towing an inflatable float from shore attach a strobe to the flag pole to warn away boaters. Dive boat captains attach a strobe to the anchor line just below the surface, to give divers something to home in on. It's amazing how far the flash of a strobe can be seen through the water. On a shallow dive it may be visible from the farthest point of the wreck. In deep water and poor visibility, a strobe should be attached to the base of the anchor line: not right on the bottom, where it's flashes might be blocked from view by intervening wreckage, but high enough above the structure so it has an unobstructed path to the horizon. A strobe set in this fashion can be retrieved when the grapnel is pulled.

Another useful marker is the chemical light, or glow stick. This handy device is a plastic tube filled with two reactive chemicals which glow when mixed. The tube is the size of a Havana cigar. It is activated by holding the stogie in both hands and bending it enough to crack the barrier in the middle that separates the chemicals. Depending upon the ingredients, the stick will glow with a variety of colors, although green is used most commonly under water because it is visible the farthest.

The glow stick is not nearly as bright as a strobe and has nowhere near the penetrating power. Nevertheless, it is often attached to the anchor line or boat ladder; it gives a larger glow when placed inside a translucent plastic jug like a one gallon milk container. Buddy teams tie glow sticks to their tank valves so they can keep an eye on each other. On a clear night you can look down through the water and delight in the ethereal effect of fuzzy green chemical blobs darting to and fro while the beams of dive lights carve crazy zigzag patterns across the bottom.

More eerie than that is the bioluminescent display of plankton. Some forms of planktonic life are natural chemical glow sticks, containing within their bodies certain chemicals that react with sea water. When the delicate outer tissues are torn, sea water contacts the chemicals, and a sparkling green point of light is emitted. This is often visible on the beach at night, when waves throw the little creatures out of the sea; the breakers glow with the emitted light, and if you walk through the receding water or tramp below the high tide mark at low tide, your footprints will glow from the chemicals released by the crushed bodies. The wakes of motor boats driving through vast stretches of floating plankton will glow in a similar fashion, as the propeller blades churn the microscopic chemical factories into a radiant green soup.

Under water, you might not be aware of the presence of bioluminescent plankton if you use an exceedingly powerful light: the intense illumination will outshine the faint chemical glow. Not unless you turn off the lights can you enjoy the effect of the myriads of protozoans twinkling like stars on a midsummer night, or flashing like meteors through the autumn sky. A shipwreck can become a surrealistic landscape as each point of contact is limned in the afterglow; your buddy might look like a child's connect-the-dot drawing.

Usually, bioluminescent plankton drifts with the current in the warmer water close to the surface. For a truly "enlightening" experience, press the lens of your light against your chest during your ascent; as you pass through the thermocline you might be rewarded by a splendid, scintillating show that will put to shame a stroll through nighttime Las Vegas or a Fourth of July fireworks display.

To put the record straight: you might hear people use the word "phosphorescence" to describe what is in actuality "bioluminescence." Technically, phosphorescence refers to the emission of light due to the irradiation of phosphorus. Bioluminescence is light (luminescence) emitted by life (bio), and is caused by the oxidation of any one of a large number of enzymes collectively called luciferin. It has nothing to do with phosphorus.

Never turn your light off under water. The two times a bulb is most likely to fail are when it is turned either on or off: when the filament either expands as it is heated electrically or contracts as it cools. That's why you should press your light against your chest to experience the effect of darkness.

The corollary to night diving is "turbid water" diving, or what is less euphemistically known as "bad visibility" diving. Whereas night diving implies diving in water that is clear but dark, with visibility possibly extending as far as the beam of a light will reach, turbid water diving involves diving in water in which so much sediment or particulate matter has been stirred up that visibility may be in the order of from a few feet to several inches, or even zero. I have been on daytime dives that were so dark that with my gauges touching my mask and my light touching my gauges, I could not make out any of the readings. This is what we call a "braille" dive; team diving is called a "group grope."

Although night diving and turbid water diving are generally lumped together as a single entity, with the connecting thread being darkness or the total lack of ambient light, there is very little similarity between the two other than the need for artificial illumination. The techniques of diving at night are essentially the same as those used for diving during the day, with the major exceptions being the addition of strobes, glow sticks, and dive lights, and perhaps a little extra caution and a shortened tether. (Shore.divers might want to leave a gas lantern on the beach in order to guide them back to the point of entry.) But when the visibility is truly awful or even nonexistent, strobes and lights don't solve the problems.

This brings us back to the line reels left at the end of the previous chapter. If you dive under circumstances in which returning to the anchor line is desirable, establishing a way-back line is the only way to ensure your return through black water or near-zero visibility. The trick is to hold a light and a

Jon Hulburt found this object in two feet of visibility. Turn the page to see what it is.

reel in the same hand, so the other hand free to spool yourself in. Try it under good conditions before you do it for "real."

If the position of the grapnel is precarious and it looks as if it might pull free, do not use secondary tie-offs. If you do, a freed grapnel moving away from you will quickly snap the line. Instead, tie your line high off the bottom (at the top of the chain, for instance) and keep it clear of all obstructions. That way, if the grapnel does pull out you'll know it as soon as the reel begins to spin wildly even though you're not moving. Hightail it back immediately by rising off the bottom to avoid slamming into wreckage or getting snared momentarily as you are pulled along by the line. If you can't keep up the speed by kicking your fins and reeling yourself in, dole out line until you are completely off the wreck and away from all possible snags; then pull yourself hand over hand to the madly dragging grapnel. If the line snaps, make a controlled ascent to the surface and keep swimming downcurrent toward the boat, shining your light as a beacon so the captain knows where you are.

Alternatively, tie your line to a solid piece of wreckage instead of to the anchor chain. Then, if the grapnel is gone when you get back to where it was, you can use the line to help control your ascent to the surface, and to permit you to maintain station over the wreck site: a position known to the captain and to which he can easily return. (See Chapter 12 for further advice in that regard.)

Whereas a line reel may be an added comfort to have on a night dive, it might be your *only* comfort in truly turbid water, when total disorientation is likely to occur within arm's reach of the anchor line. Again, it pays to become proficient with line-reel use before you try it under full blackout conditions.

By now you might be asking, "Why bother diving in such bad vis?" And I might reply, "Because you can't swim in a desert." For those of you who just said "Huhn?" I will elaborate. You can't swim in a desert because there's no water there. By the same token, the wreck you want to dive may not have had the good fortune to sink in a clear crystalline sea. Since you can't take clarity to the wreck, you take the conditions as they come. Notice that I never said I *liked* diving in coal black water where I couldn't see my hand in front of my face, I just said that I've done it.

It pays to be prepared for bad visibility because you can encounter it

It's the conning tower telegraph from the U.S. submarine *S-49*, sunk in the Patuxent River, in Maryland.

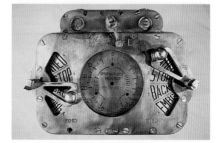

when you least expect it. Many are the times that I've gone down through clear surface water, only to discover that the colder, denser water below the thermoline was socked in with fine silt or large particulate matter, or that deep-water ground swells had stirred the bottom into the consistency of egg drop soup. More than once I had no inkling that I was approaching the wreck until I felt the anchor chain in my hand or banged my mask on the wreckage. At that point the choices are to scrub the dive or to accept the fact that it's going to be a different experience from the one that was planned. Try to see the positive side of things—you may as well, because you can't see anything else.

I've also found things in bad visibility that I never would have found otherwise, because I would have been looking into the distance instead of at what was immediately beneath my fins. When the visibility is great, reconnoiter the wreck and take an overview tour around the perimeter, and study the whole in a broad perspective that will establish in your mind a lasting image of the layout; when it's bad, switch to macro mode and zoom in on the tiny things you've never noticed before: colorful marine encrustations, the tentacles of sea anemones, baby coral polyps, diminutive tropical fish, hermit crabs walking with their shells, and all the things that live on the bottom that you're always too busy to watch. I have even discovered partially buried artifacts that I never would have spotted had I been swimming by at a height of several feet. Sometimes it pays to put your head in the sand. Now's the time to dig for goodies.

The greenhorn might think that a brighter light will dispel the gloom of night, and he'd be correct. But if he thinks that higher wattage will pierce through silt and mud and drifting hordes of plankton, backscatter will come as a big surprise. (See Chapter 8 for a description of backscatter.) To understand the principle involved, ask yourself this question: how bright a light do you need to have in order to see through walls?

Particulate matter, whether it consists of tiny particles of silt or large masses of plankton, is a wall. Instead of existing on a plane, however, it extends in three dimensions—it has width in addition to length and breadth. All water contains silt in suspension—that's why Caribbean visibility is limited to a couple hundred feet instead of being as clear as air; it's analogous to underwater fog.

In extremely turbid water the particles are more numerous and closer together. It's as if one piece of the wall is a foot away, another piece is two feet away, and yet another piece is three feet away: all multiplied by thousands or millions of times. The end result is a tangible barrier impenetrable by light. Certainly, a powerful dive light may enable you to see several feet instead of not at all, but there is no radiation in the visual field of the electromagnetic spectrum that can pass through solid matter.

What you see is what you get. And sometimes what you don't see.

Lasting Images:
Basic Wreck Photography

Underwater photography is one of the most frustrating endeavors you will ever undertake. It can also be one of the most rewarding.

The tyro who gawks at the colorful marine life pictures and majestic shipwreck photos that grace the pages of books and magazines and that flash on the screens at underwater symposiums, and imagines that he can easily produce results of equal quality, is bound to be disappointed. Successful underwater photography requires a considerable investment in money and time, and takes a great deal of dedication.

Contrary to popular belief, photography is not fun. It is work—very hard work. Underwater photography is even harder. It's work because instead of simply exploring a wreck or enjoying the thrill of the deep, you've got to drag around a camera system and concentrate on creating art: framing shots, looking for suitable subject matter, double-checking exposures, and setting proper apertures and f-stops. This is no simple process of point and shoot. You give up the broader picture, and see the wreck only through an eyepiece or framer.

The fun part comes when you get photographs you are proud of. But the road between visualizing images and capturing them on film is a long one, fraught with pitfalls along the way. Let's talk about the frustrations first. Then, if you still want to shoot pictures under water, read on.

The frustration usually begins when you compare prices in the stores and catalogues and find out how much an underwater camera costs. It gets worse when you discover that the quoted prices often include only the camera body (read the small print) and not the lens. Then, after you dig deep into your pockets and pay what may seem like an exorbitant amount, you learn that the camera and lens will not take underwater pictures in vivid color like you see in the slicks. For that you've got to buy a strobe light that costs more than the camera. Furthermore, the strobe requires some rather expensive accessories: a base plate, a ball-joint bracket, and an extension arm (to physically mount the strobe to the camera), and a waterproof electric cord to connect the flash tube assembly to the shutter release mechanism.

But it doesn't end there. It never ends. Now you've got to buy film and pay for processing. You may want to take an introductory photo course. Then you've got to visit exotic locations in order to practice what you've learned: this means air fares, hotel fees, gear rental, boat charter expenses, and miscellaneous charges like meals, tips, and local transportation. After you've taken a second mortgage on your home to cover the gross expenditures, you find that you need more room to stow your gear, more drawers and shelves to store negatives, prints, and slides, and more walls on which to hang those favorites that you've framed. Obviously, all this paraphernalia demands a larger house.

Seriously, though, after the initial outlay comes a conspiracy of neverending bills for what the neophyte seldom reckons on: maintenance and repair. At times, disbursements may be made with such irritating regularity that underwater photography seems more like extortionism than an art or business venture. Let me tell you how it is with me.

The Nikonos is probably the most widely used self-contained underwater camera in the world today. It has been around for decades in various models beginning with the Roman numeral I and is now up to V. In order for me to keep one Nikonos working in the water, I have to own three of them. At any one time I might have one Nikonos being repaired, one in use, and one as a backup because the chances are that the one I'm using will malfunction before the third one comes back from the Nikon repair shop. Another way of stating the situation is that each of my Nikonos cameras breaks down and must be sent out for repair at least once a year, which, since I own three of them, means that I get only four months use out of each camera before it bites the silt. Couple this with the fact that the average repair time is four months, and you can understand why I have to own three of them. This is not only frustrating, it's expensive.

The minimum repair cost, including two-way shipping and Nikon's break-down fee, is about $100; and that's without parts. Granted that my cameras get more use than most, it still doesn't alter the fact that I never get more than a few dozen rolls of film through a camera—even a brand new camera—before it goes kaput. By contrast, for topside photos I use a Canon

These close-up photos were taken in less than three feet of visibility. I shot 400 ASA film, used a 15-mm lens placed less than one foot from the subject, detached the strobe and held it above the subject at nearly ninety degrees to the line of sight, and took several manual exposures.

A-1, and in all the years I've owned it I have *never* had to have it repaired—and I run about ten times as much film through it as I do for each Nikonos.

The point of this discussion is to introduce you to the lack of dependability in underwater camera systems. It gets worse when you add strobe lights and synch cords. If every part of the system doesn't work perfectly and at the same time—well, you get the picture. Or do you?

And I'm just talking about normal wear and tear, not the major disaster of flooding. With a minor fresh water leak a camera may be repairable, but costly. With a major salt water leak your best bet is usually to throw the camera away and buy a new one. (I trash about one Nikonos a year.) There are numerous gradations in between, running from "cost effective" to "not economically feasible." On a Nikonos V, for example, a few drops of water entering the body through the shaft orifice of the rewind knob may wipe out the electronics because, for some reason, they are not encapsulated in plastic. The camera can be fixed, but at what I call a "creeping cost." That is, it's slightly cheaper than buying a new camera, so you ante up. Then, a few months down the line you discover that internal spot rusting that was not "caught" during the break-down and repair process initiated a series of insidious, one-at-a-time mechanical faults that results in the camera becoming a repair shop junkie: returning again and again for a "fix" that is never enough to satisfy the hunger of seemingly unrelated malfunctions. This becomes a nightmarish game of what's-wrong-with-it-this-time; the camera becomes a shutterbug yo-yo, bouncing around the world and making the shipping companies rich, but tied to you with a string that has a strangle hold on your resources. Each repair job offers false hope of a permanent cure, and is as draining on your emotions as it is on your bank account. The result is a long, lingering death for both the camera and the would-be photographer. Heart disease, at least, is predictable.

So why don't you ever read about the afflictions of "camera repair syndrome"? For the same reason that tobacco companies don't advertise the health hazards of cigarette smoking—it doesn't help sell the product. Notwithstanding these graphic observations, I do not intend to repudiate the potential joys of underwater photography. My purpose in disclosing these hidden truths is twofold: to warn those who are about to commit themselves with ingenuous glee, and to commiserate with those who have already been committed, and who thought paranoiacly that their persecution was unique.

For insights into care and preventive maintenance for photographic equipment, follow the procedures described for dive lights in the previous chapter. Since strobes are also powered by either disposable batteries or rechargeable ni-cads, those instructions and suggestions apply as well. The only extra bit of caution is this: after salt water immersion, the sync cord should be disconnected after rinsing, and all exposed threads—male and female—should be cleaned with a toothbrush. If the sync cord is left connected for too long, especially if it is screwed into the camera or strobe during storage, corrosion products and pieces of grit that accumulate on the threads will "freeze" the plug to the bulkhead. You may need pliers to unscrew the cord connector, or you may have to send the affected unit to a repair facility for separation.

Nikonos users beware: even when you shoot without a strobe, after salt water immersion you should unscrew the strobe port plug and clean all the threads. This is necessary because the o-ring is situated on the inboard side of the plug instead of the outboard side. The pressure of depth forces water around the threads, where it stays until it evaporates. The mated aluminum surfaces oxidize to the point where they cannot be disunited. The plug then has to be drilled out—but carefully, so the bit does not destroy the flash socket pins. (You won't find this tidbit in the instruction manual.) Ironically, the battery port plug is designed with the o-ring on the outboard side. Users of the Sea & Sea Motormarine need not worry, because the plug and camera body threads are made of non-corrosive plastic.

An alternative to buying a so-called "amphibious" camera is to purchase a waterproof housing for the topside camera that you already own. At first thought this seems like a less expensive investment, but on second thought you will discover that the cost of a good housing rivals the sale price of an amphibious camera. The main advantage that I've come to find is that I can now take my reliable Canon A-1 under water, and use the power winder as well. And since I already have a full array of lenses for the Canon, it obviates having to buy a duplicate set of lenses for the amphibious camera: a significant savings that goes beyond the comparison between the cost of a housing and an amphibious camera body. Not only that, but the use of a dome port increases the coverage angle of a 20-mm land lens to that of a 15-mm underwater lens— at less than half the cost of a Nikonos 15-mm.

The major objection to housed cameras that most people voice is the bulk. Yet, although it is true that a Nikonos or Motormarine is only slightly larger than a pack of cigarettes while a housing is nearly the size of a bread box, there is no overall difference in the size of the *system* when a strobe and its accessories are adjoined. Add the convenience of single lens reflex (SLR), which does away with the parallax error problem, and a housed camera becomes more attractive.

Here are some tips for housed camera users. Because light can reach the film through the eyepiece of an autoexposure SLR and alter the exposure, be careful to block unwanted ambient light by pressing your mask against the back of the housing when you release the shutter. A wad of toilet paper in the bottom of a clear plastic camera housing will not only soak up water from minor leaks, but the change in color will warn you that all is not well. Before screwing the strobe cord connector to the camera housing bulkhead connector, apply copious amounts of dielectric silicone grease to the electrical pins and sockets: this ensures the transmission of current needed to trigger the strobe, and prevents corrosion. Generally, the strobe will not fire if the connectors are ungreased. If you travel from a high elevation to sea level (say, from Denver to the Caribbean), either disassemble the housing or remove the dome port; otherwise, the increased pressure from the loss of altitude might exert so much force that you won't be able to open the housing when you reach your destination. In the reverse case—from a lower elevation to a higher—the lack of pressure will place a strain on the housing clamps. Do *not* remove the o-ring as some people suggest; it's too easy to forget to put it back in and cause the housing to flood.

It is beyond the scope of this chapter to discuss everything there is to know about underwater photography. Entire volumes exist on the subject, replete with definitions of f-stops, apertures, depth of field, parallax error correction, and many other esoteric terms and techniques that relate to taking pictures in the underwater realm. You should refer to them for basic information. I could not do justice to the matter without expanding this section to the length of a book. Instead, my purpose is to concentrate on providing insights that are left out of the manuals and magazine articles (as in the foregoing passages) and to describe the applications of photography as they relate specifically to shipwrecks.

Let me introduce you to Gary's Three Rules of Photography: always, always, always take your camera. The one time you leave your camera behind is the time you'll find that perfectly intact artifact lying exposed in the sand; and by the time you get back to it on the next dive, it may already have been recovered.

As with any rule, there are exceptions, the main one being adverse environmental conditions: seas so rough or current so strong that dragging a camera through the water might prove injurious to either the photographer or his equipment. Other extenuating circumstances might apply; you be the judge.

One excuse that doesn't hold water, so to speak, is what appears from topside to be limited visibility. I have seen surface water that looked like pea soup, but found clear water underneath: a phenomenon I call "plankton gloom" that fades with depth or terminates abruptly at the thermocline. The bottom will be dark because sunlight is blocked by the intervening plankton (not the day for available light photography) yet otherwise worthwhile for strobe work.

If you get to the wreck and find it almost totally socked in, shorten the focus and do close-up photography. If you suspect beforehand that the

Shipwrecks offer unique opportunities for macro-photography. Above: Clips of bullets for World-War-One-era Springfield rifles; the brass safety pins probably held the cloth wrapping together. Right: A ceiling fixture with the glass globe intact; the light is inverted because the wreck is upside down.

visibility on the bottom might be bad, use the opportunity to catch up on your macro work; keep your close-up lens and extension tubes with you at all times. Motormarine users can take this suggestion quite literally, since lenses are interchangeable under water. (Motormarine lenses and camera bodies are independently waterproof. With the Nikonos system, if the lens leaks the camera gets flooded as well.) Under circumstances of poor visibility, be mindful of particulate matter and the backscatter effect.

Suspended particles reflect light directly back to the source. This is why you can't see when you use your car's high beams in a fog; switch to low beams and the roadway reappears. The same mechanism operates in the water when silt or microplankton bounces back light from a dive light or strobe. If the path of the beam closely parallels your line of sight or is nearly perpendicular to the film plane, you and Kodak see the cumulative reflections as a burst of bright luminous dots. This optical effect is called "backscatter." It ruins more underwater photos than all other causes combined.

There are ways to avoid or at least diminish backscatter, there are ways to save pictures that are marginally backscattered, and there are uses for photos that are horribly backscattered. Using the headlight analogy from above, it becomes apparent under water that the wider the angle between the beam of light and line of sight, the more that backscatter is reduced. This is why strobe attachment arms are so long. Sometimes, you might even want to detach the strobe and hold it out to the side or above the subject, in order to get near right-angle coverage.

The shorter the distance between the camera and the subject, the fewer light-reflecting particles there will be: thus the rationale for close-up or macro photography. Furthermore, by being close to the subject less light is necessary to properly illuminate it. Less light means less reflection. Use faster film and select a lower power setting.

More important to wreck-diving photographers is understanding that shipwrecks generate their own particulates: flakes of rust, silt from rotting wood and other organic materials, the residue from deteriorating cargo and miscellaneous ship components, dead and disintegrating marine encrustations, and excretions from fish. Due to the complicated structure of collapsing hulls and the innumerable interstices provided by the degradation of wooden beams and metal plates, these particulates become trapped: like tiny time bombs waiting to explode at the slightest fin kick. In addition, silt in suspension and moving with the current is captured by the wreck. Throw in deceased plankton raining down from above, and it is easy to see (or not to see) that wrecks are predisposed to bad visibility.

If you plan to take pictures, get in the water first before the masses descend upon the wreck and stir it up. Practice anti-silting techniques: maintain buoyancy control so you can hover above the bottom, and approach your subject with mild flutter kicks that do not create the swirls of silt that will overtake you like dust on an unpaved road. Hold the strobe at various angles and take lots of pictures—film is cheap compared to the cost of the dive. Later, at home, you can sort out the good shots and trash the ones that didn't turn out. And *never* let anyone see the bad ones.

The cardinal rule of all photography is to keep culling your stock: show off only the pictures worth showing. If you think this is wasteful, know that in the business if a photographer gets from a roll of film a single image that is publishable or presentable, the shoot is considered a success; two is a gift from

Proper wreck photo technique: good buoyancy control enables the photographer to hover above the wreck; by keeping his legs bent his fins do not kick up the ever-present bottom sediment; and the strobe is detached in order to reduce backscatter.

Simple cropping not only eliminates the lighted particles on the left side of the picture, but it deletes the dead space on either side of the subject. By changing the format from horizontal to vertical the composition has been improved and an entire new image has been created, one that accentuates the deadeye while surrounding it with fish: a secondary theme.

heaven. I have shot as many as three rolls of film on a single subject, in order to ensure getting one picture that was marketable.

Some photos that are flawed by backscatter can be saved. Typically, because the strobe is mounted on the left side of the camera, the annoying luminous dots occupy only the left side of the picture. If you have prints made from negatives or slides, crop out the unsightly portion with a scissors, or frame the print with a decorative mat. If you have enlargements made, trim the print in proportion so the viewer cannot tell that the picture is only part of a much larger image.

For slide presentation purposes, buy masks of various sizes and shapes (circles, ovals, etc.) or cut your own from black masking paper. Inserting these cutouts into the mount creates an artistic effect that conceals the imperfections. If your slides are in cardboard mounts, you can always buy plastic mounts and remount the slides that you want masked.

Cropping can also add balance and improve composition in a photograph that is otherwise off-centered or poorly framed, such as a picture snapped on the drift in a fast current or while being battered by a monster surge. Fish are uncooperative and take direction poorly, often passing in and out of the frame while you compose your image. The result may be a disturbing splash of light off the scales of a fish that darted close in front of the camera just as the strobe discharged. Sometimes the fish can be eliminated by masking or cropping; alternatively, the bright spots can be corrected in the printing process, by burning in the areas that are too light, but this requires professional treatment or advanced darkroom technique. Another problem is caused by fish that are curious and that school haphazardly near the subject, giving an unorganized appearance to the final image. These are reasons to take a series of photos instead of just one quick snapshot.

Some pictures—even bad pictures—may have informational value that warrants saving them. Photographs taken for survey purposes can help piece together the layout of a wreck despite poor quality. You might also want shots of artifacts or structural members for study or comparison. It's okay to hang on to pictures like these, just don't display them.

Many people are convinced by clever advertising claims that a camera equipped with through-the-lens metering (TTL) guarantees perfect exposure with every shot. In strobe-assisted wreck photography this is not necessarily the case.

I use automatic exposure on my topside camera almost all the time, and it works nearly perfectly. Automatic exposure mode works equally as well under water for available light photography; it not only takes the guesswork out of metering, it allows for split-second adjustments right up to the time the shutter is tripped. The camera does this by reading the amount of light that reaches a light-sensitive cell located within the body, and correlating the information with the ASA setting (film speed). When the proper exposure is reached, the shutter closes.

In conjunction with a strobe, TTL utilizes the camera's automatic exposure meter to adjust the amount of light by regulating the duration of the flash. The fallacy of the system is that it works well only on a relatively flat plane, or when there are no dark areas surrounding the subject. This is because the meter measures the *average* amount of light reaching the photocell. During the day, the amount of sunlight on the wreck is pretty much the same all over (except, of course, in the shadows); this is because the light source is ninety-three million miles away, and a couple of feet more or less doesn't make much difference.

With a strobe as a light source, however, the inverse square law comes into effect. Simply stated, doubling the distance that light travels reduces its intensity to one-quarter; triple the distance, and the intensity is reduced to one-ninth. Thus the amount of light falling on a subject at four feet from the strobe is one quarter the amount of light falling on a subject that is two feet from the strobe. This means that if a diver poses in front of a vertical hull plate, TTL will shut down the strobe when the photocell receives the amount of light *averaged* from the two subjects: diver and hull plate. The end result is that the diver is overexposed (because he is closer to the strobe), the hull plate is underexposed (because it is farther away from the strobe), and the picture is worthless. The same consequence occurs when TTL is used to photograph a diver alone in dark water; the surrounding black space reflects no light at all, so TTL keeps the strobe going until the amount of light reflected off the diver equals the amount of light the photocell expects to receive from all four corners of the frame. The outcome is a totally washed out diver set against a black background. The problem is further exacerbated by particulate matter (even if invisible to the naked eye) absorbing more light as the strobe-to-subject distance increases.

In the first example, it is patently impossible with a single strobe to take the picture with equal exposure on both the diver and the hull plate—with or

Two tips to the wise: Notice that the photographer has placed an automobile tire inner tube strip around his strobe in order to prevent fishing line from slipping under the latch and flipping it open. The picture above shows how the inner tube should be placed in order to prevent a catastrophe. It might not look professional, but it works. Left: I spray my lens covers fluorescent red; the paint chips easily and looks unsightly, but it serves its purpose well: it reminds me to remove the lens cover before entering the water. If I leave it on by mistake, a diver whose picture I am taking is more likely to notice a bright red disk than the flat black of the lens cover, and can alert me to the error of my ways. Many a roll of film has been shot with the lens cover in place.

without TTL. The photographer must decide which subject matter is more important, and expose manually for the distance to that subject. In the second example, the proper exposure can be made but only by shooting manually. In a further example, white coral growing next to red coral will also cheat the meter and throw off the exposure, because each color reflects a different amount of light.

This is not to say that TTL does not have its place. In macro photography, when the framer is pressed against a flat surface so that all points within the field of view are equidistant from the light source, remarkable results can be achieved *as long as* the strobe is not placed at too divergent an angle, where more of is light reaches one side of the subject than reaches the opposite side. TTL is not foolproof—it requires some knowledge of optics and a little experience in the water. If you want to learn from your mistakes, take a slate with you and keep a photo log: note each shot with angles, distances, and subject matter.

Wreck photographers should be wary of fishing line: monofilament is clear and difficult to see. It snags all too readily on a camera system's knobs, levers, and latches, not only entangling the unit but possibly damaging the controls or causing the camera or strobe to flood. The anchor line can cause the same effect. Because the latches on my strobe do not have safety locks, I cut a fat strip from an inner tube and place it around the body of the strobe and over the latches; this prevents monofilament from sliding under the latches and popping them open (and consequently flooding the strobe). Another method is to make a neoprene sleeve from scrap wetsuit material and pull it over the body of the strobe; not only does this cover the latches, it protects the strobe from scratches and adds buoyancy to the system.

To help ensure against loss, attach an inflatable marker buoy to your camera system. In case of an emergency, activate the CO_2 cartridge and send the camera to the surface for later retrieval.

Of all the wreck diving disciplines, photography is undoubtedly the most difficult to master. Its frustrations are endless, but its rewards are unlimited. However, before tackling a pursuit that requires such strict attention to detail, it is first necessary to be fully at ease with the underwater environment, with your gear, and with yourself. Once this is achieved, give full vent to your artistic talents and see where they may take you. Satisfaction can be yours at the snap of a finger.

Camera Repairs

Southern Nikonos Service Center
9459 Kempwood
Houston, TX 77080
(713) 462-5436

Underwater Photo-Tech
16 Manning St., Suite 104
Derry, NH 03038
(603) 432-1997
(FAX) 432-4702

My buddy shines a light through the eye of the ram's head figurehead on the *Sandusky*.

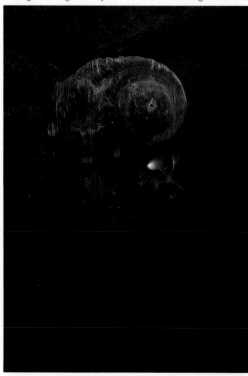

Shape Up or Shipwreck Out

Too often have I heard people claim that they could never learn to dive because they were too old, out of shape, afraid of the water, or a number of other excuses that were part of the litany of a negative outlook on life: predictive justifications for failure that ensure the lack of successful endeavor. In contradiction to such a posture, I believe that anyone who can walk and chew gum can learn to dive, and that the major prerequisites are motivation and drive. Between these two points of view lie genuine emotional and physiological factors that determine a person's degree of comfort in the water and the level of expertise that a person can achieve. Insurmountable among them is genetic heritage.

Practically everyone can cover a mile on his own two feet; very few have the athletic ability to run the distance in four minutes. Meaning? Those who are not so fast should travel at their own pace and stop to smell the flowers along the way. Like most pastimes, wreck diving is not a competition, but an activity whose satisfaction comes from within. You compete only against nature or yourself: striving for your personal best. No one loses in the game of life except those who choose not to engage life to the fullest. All are winners who pursue firm objectives, whether it be exploring a newly discovered wreck, taking that once-in-a-lifetime photograph, encountering rare marine animals, recovering a hard-to-get porthole, or savoring the awe and beauty of an artificial reef. Challenge and fulfillment can be yours.

Yet it must be acknowledged that, due to the extreme physical nature of diving and to the psychological stresses induced by submersion in a non-life-sustaining medium, where performance is dependent upon handling of equipment, personal resourcefulness, and the proper response to stimuli, it pays to be physically fit and mentally prepared to meet the multitudinous situations that arise during the course of the ordinary dive, and that may arise during an extraordinary dive. Confidence in your ability to conquer adverse environmental conditions is a self-perpetuating modality that serves to reduce pre-dive stress and conserve energy while enabling you to exert your full strength and stamina when needed. This makes diving safer, more enjoyable, and, in the long run, more rewarding.

Individual limitations aside, stress induced by the imminent failure of a diver to resist the physical strain of cold, current, wave action, or weight of gear can best be avoided or overcome by preconditioning: a light regimen of muscle toning and aerobic exercise such as that recommended by doctors for general good health. I don't mean bench-pressing five hundred pounds or running Boston marathon, just in-house calisthenics or regular visits to the local Y, and jogging around the block. The difference that a steady workout program can make is astounding. For example, a person in poor physical condition surfacing away from the boat in a stiff current may become stressed immediately to the point of panic, whereas a person who is fit and knows it merely ducks his head in the water and kicks for all he is worth.

Lest you fall prey to the "gorilla diver" fallacy, pure strength is not worth as much in wreck-diving circumstances as endurance. In the example cited above, the hulking brute with haunches built like steel pylons is not as likely to make it back to the boat as the slim but trim swimmer who can't lift as much weight with his legs but who can keep on kicking and whose strength does not run out after a short spurt of speed. Another way of stating it is that big people are often strong due to sheer muscle mass, but lack endurance.

The point I want to make is that weight training is not as important as increasing cardiovascular efficiency. The only times pure strength is useful is for climbing boat ladders or making shore entries and exits, and transporting gear topside. Otherwise, stamina rather than strength will serve you better in the water.

For the person who wants to undertake wreck-diving as a steady activity and who wants to become proficient at it without overstressing himself, maintaining a schedule of aerobic exercise is the best form of conditioning. This is because so much of diving involves proper breathing. Swimming and jogging are the best forms of cardiovascular exercise, with in-place aerobics coming in third. Swimming is less traumatic on the joints than jogging, and also exercises the long thigh muscles that are used in the full fin kick. If a pool is not available, or if you don't like swimming or find pool hours inconvenient, jogging may be more suitable since you can do it most anytime. For those who live in climates with cold winters and frequent snow, or who can't stand the heat of summer, videotaped workout programs will guide you through the motions in the comfort of your air-conditioned home.

(As a rule, I never jog *after* a dive until the next day. I figure that when my body's level of saturated nitrogen is high, increased circulation may lead to symptoms of the bends. Besides, the work of diving is enough exercise for one day.)

It is equally important to practice deep breathing exercises while diving, where long, deep breaths make more efficient use of the air in a tank than short, shallow breaths. This is because not all inhaled air reaches the lungs, where the oxygen is extracted; some air is drawn only as far as the trachea, and is then exhaled without being utilized by the body. On the surface, air is available in unlimited quantities. But in scuba, each cylinder contains only a specified amount of compressed air, so a wasteful breathing pattern can shorten a dive or reduce reserves. Concentrate on developing a good breathing rate, practice it all the time under water, and be consciously aware of each breath; eventually, it will become second nature.

Some people get painful leg cramps during a dive. Despite what your mother may have told you about entering the water less than an hour after eating, most cramps occur because the muscles are being overstrained. To prevent cramps and fatigue, divers often resort to finning through the water

Jogging is great cardiovascular exercise, bicycling is a good way to strengthen the legs without the stress of jogging. In the winter, when snow blankets the jogging trail, I enjoy the solitude of cross-country skiing. A constant regimen of exercise will help increase your SAC rate (surface air consumption rate), and make diving less physically stressful.

with a "bicycle" or "peddling" kick: a bent leg kick in which the upper leg is practically stationary and most of the propelling force is supplied by the calves. This is equivalent to walking with steps in which the heel of one foot touches the big toe of the other: you'll get where you're going, but at a snail's pace compared to the more comfortable and efficient stride that utilizes the full length of your legs. In the water, a straight leg kick derives its strength from the large thigh muscles which, in everyday use, may be well conditioned for walking and jogging, but not for fin kicking. To exercise these muscles for diving, lie on your back, raise your legs to a forty-five degree angle, and swing them up and down.

As with any exercise, there's no reason to be fanatical about performance. Because so many factors other than muscle tone are involved (for example: rest, diet, and level of hydration) you may tire more easily one day than the previous, leading you to believe that you are degenerating. Conditioning is something that comes with longterm routine, and acclimating your body to work in the underwater environment takes time. You probably won't notice a significant increase in strength or endurance on a day-to-day basis, and you may never be fully aware of how well you have adapted your muscles to handling your equipment and to overcoming the rigors of the world underwater.

Don't get frustrated like my good friend Evie Dudas, who complained that on one dive in fast current the fish swam effortlessly among the reefs, while she struggled to keep from being swept away. I said, "Don't feel bad. If those fish were on land you could outrun them."

You should also arrive at the wreck site fully rested. This means no late-night activities before a dive. Get a good night's sleep so you wake up refreshed and can look forward to the day's adventure. Don't be afraid to nap on the boat on the way out to the wreck; and if you feel the need, catch a little shuteye during your surface interval. This will help increase your alertness on the bottom where it counts.

The best way to gauge your efficiency level in the water is to calculate your air consumption after each dive. Use the formula from your basic certification course (the correlation between depth and time), but keep in mind that wreck-diving adds variables that may offset a precise computation: current, surge, waves, and distance of exploration. Concentrate on breathing and kicking efficiently. Continued monitoring will give you an up-to-date progress report on your condition, and point out where you need to focus your training regimen.

Familiarity with your gear and confidence in your ability to work effectively will increase your comfort in the water, and put you in a mental state in which you will be more likely to surmount the obstacles of nature, be better able to cope with stress, and be more fully aware of your surroundings. In the grand scheme of things, that is one of wreck-diving's greatest rewards.

In the final analysis, the best exercises for keeping in shape are forearm presses and bicep extension—practiced by holding your utensils down on the plate and pushing yourself away from the table.

Backpacking is another great way to keep fit. And the scenery is magnificent.

Rx for Diving

They say there are four stages to seasickness: first you think you're going to throw up; next you reach the point where you can't throw up any more and just go through the motions; then you think you're going to die; finally, you're afraid you won't die. As one who continues to suffer through all four stages despite my years at sea, I am here to say that although severe and neverending nausea does not necessarily threaten the prolongation of life, it is an unforgettable experience that can ruin your whole afternoon.

"But don't you ever get used to it?" the novice is wont to ask. Let me put it this way: do mountain climbers ever get used to falling off cliffs and slamming into the rocks below? If you are prone to seasickness, chances are you'll be prone to it for life. By this I do not mean to imply that every boat ride will be an upchucking trip to hell, but that you'll probably always feel a bit queasy even on the best of days. Sometimes, anticipation alone is enough to push you over the edge. If the flags are flying straight out, if the wind is howling about your ears, if the waves are breaking in the marina, these are little indications that it might be a good day for you to keep your feet on the ground.

One time, while climbing an ice-covered mountain in the dead of winter, I had a flashback that helped to mitigate my discomfort from the cold. Caught near the summit in seventy-mile-per-hour winds that were gusting into the nineties, with the temperature down to twelve below zero, and with snow stinging my face and beating into my eyes, I lay on my side and carved out a snow-cave with my ice-axe. I spent the night shivering despite heavy-duty longjohns, a sweater, a parka, and a down sleeping bag. The one thought that kept recurring, and that maintained my spirits throughout the frigid hours of darkness, was that *mountains don't move.* I may have been freezing, but that was better than being seasick.

If you have never been to sea, it is impossible to predict whether you will get sick your first time on a boat. Nor is it possible to predict, even though you don't get sick the first time, that you won't get sick the next. I have been out in some awful storms, in waves from ten to fifteen feet in height, and have felt fine; I have also puked my guts out on round-bottomed boats that rolled with a corkscrew motion in a gentle swell on an otherwise calm and placid day. There is only one sure-fire prevention for what the French call *mal de mere:* stay home.

Mild cases may come and go. Sometimes, I feel much better once I vomit, as if I've gotten the "bug" out of my system; I may go on for the rest of the day without feeling the least bit queasy, and might even get my appetite back. At other times, though, I get horrible headaches in which the dimmest light sends stabbing pains through my eyeballs, like ice picks into my brain.

Mountains can be cold, but at least they don't move.

Worst of all is getting sick under water. Like the time that my friend Trueman Seamans saw me surrounded by a thick dark cloud, and thought that I was shaking the mud out of a juicy artifact; he nearly gagged when he got close enough to look into my goody bag and realized that I was recycling breakfast through my regulator.

"Isn't this bad for the regulator?" you might wonder. Well, it isn't good. Large chunks of undigested food can get stuck in the mouthpiece or jam open the diaphragm; chew your food well so it comes up in fine particles. Nevertheless, it is highly recommended that you do *not* take the regulator out of your mouth while vomiting because the reflex action often induces spasms which might force you to inhale water uncontrollably. Thus the spasms may cause you to drown.

Seasickness is so commonplace on dive boats that an entire system of etiquette has evolved because of it. Vomiting is nothing to be embarrassed about; it happens to most people at one time or another. If you feel the forces of nature about to take control, head for the rail and heave to. The *lee* rail, that is—the one that is downwind; otherwise, you're likely to have the wind throw back into your face what you just disgorged from your stomach. It is considered bad form to throw up on other people or their gear, and to take food from their coolers when you've determined that they won't be eating that day. Try to avoid puking in the head because the resultant stench in the confined cabin space may cause others to join you in sympathy. If someone is sick and you are not, be considerate of his condition and do what you can to alleviate his distress: give him the best seat, get him water if he is too sick to move, and offer your jacket if he appears cold. I always carry a small bottle of mouthwash so I can rinse out the bad taste after a tummy tantrum; it helps to prevent repetitive bouts.

In the final analysis, seasickness and the degree of susceptibility are governed largely by factors beyond your control: heredity. But while you may be predisposed to suffer from the "badness of the sea," there are precautionary measures you would do well to practice.

What can you do when you've lost your lunch over the side or have gone down below to pray in the china chalice? Not a thing. Once you've got it bad, the only cure is to get off the boat. Back on dry land you'll feel better right away, as if a switch suddenly turned off the nausea notion. You'll probably be starved as well.

Some people have found that floating in the water makes them feel better: the up-and-down motion is not complicated by swinging from side to side; they may get well enough to make another dive. Then, once under water and away from the disturbing surface motion, they feel great.

To alleviate the symptoms of seasickness, or at least to prevent the dry heaves, try eating crackers, pretzels, or unbuttered bread—food that sticks to your innards. On the morning before a dive, stay away from eggs and other greasy victuals that slide around in your stomach and can easily escape through the upper hatch. If the atmosphere below deck is close and hot, stay near a porthole or some place where the ventilation is good, or get out in the open where you can inhale fresh, untainted air. On the other hand, watch out for diesel fumes. One whiff of unburned hydrocarbons is enough to send me rushing for the gunwale, so I stay away from exhaust smoke that thickens the after deck.

If you feel any hint of nausea, find a place on the boat that is dry and comfortable, and try to relax. Keep away from the bow, where the hull pounds the hardest against the waves; and keep off the flying bridge, where the swinging motion is accentuated the most. The best place to be on a boat is right in the middle and at the water line, the pivot point around which the hull pitches and rolls: this is equivalent to being at the center of a merry-go-round instead of at the perimeter. When feeling sick, some people like to lie down, others prefer to sit, some even stand. It often helps to be on your feet so you can bend your legs with the roll of the boat; this enables you to keep your head—where seasickness originates—on an even keel This is not to suggest that seasickness is a psychological ailment. Although motion sickness is all in the brain, it is not all in the mind.

The exact physiological mechanism that causes motion sickness is not clearly understood. The brain interprets incoming data through a variety of sensory devices—the five senses—and utilizes the information to establish the body's relationship to the world. Equilibrium (the balance needed to stand upright without falling over) is maintained mostly by interpreting data transmitted by the inner ears, whose semi-circular canals contain a fluid that touches nerves which then transmit signals to the brain, similar to the way in which upper and lower electrical limit switches maintain the level of water in a pool. If the inner ear fluid "sloshes" like liquid in a shaken glass, contradictory signals result in a misinterpretation of events; furthermore, nearby nerves that control involuntary functions—nausea and vomiting, among others—are stimulated erroneously. The resultant physical dysfunction is manifested by what we call seasickness in all its forms and grades.

The same mechanism can be triggered when what the brain "hears" does not correlate with what it "sees"; that is, when the fluid in the inner ear canals is steady but the eyes are disturbed by the motion of distant objects. Thus, it is possible to get sick on land while observing a boat that is being tossed about by waves, or by watching a spinning pinwheel whose surface is painted with a spiral design. In each case the brain is "tricked" by conflicting signals.

Unfortunately, closing your eyes to resolve the conflict does not make seasickness go away. But it sometimes helps to concentrate your gaze on the horizon—a stationary plane; this can "countertrick" the brain into believing what it "sees" rather than what it "hears," and might lessen the severity of seasickness symptoms.

When all else fails, try drugs. Medication that prevents motion sickness is available over-the-counter in the form of pills or capsules. There are several brands, and all of them work quite well. However, chemical derivatives that are effective for one person may not be as effective for another. Experiment with different types and find out which work best for you. If you cannot find a suitable remedy from nonprescription medicines, see if your doctor can prescribe something more effective.

In order to work at all, drug therapy must be started prior to the onset of symptoms. It takes time for the drug to get into the system, and once vomiting begins, the pills come up as fast as they go down. For dosage and treatment schedules, follow the instructions on the package. I generally take my medicine at the dock, after I've ascertained that (1) the seas are rough enough to warrant medication, and (2) the boat is going out despite the conditions. The reason for waiting until the last possible moment is not to save the cost of the pill, but to avoid being medicated unnecessarily. Most motion-sickness pills are antihistamines, and have the same undesirable side effect as the variety taken for colds: they cause drowsiness.

Too many times I've found myself so sleepy on a boat that I could hardly keep my eyes open. Putting my gear together and getting dressed for the dive was as laborious as wading through syrup. I always wake up when I hit the water, but there's that thought in the back of my mind that my reflexes are not as responsive as they should be and my level of awareness is suppressed. Therefore, it is important to take the minimum dosage needed to prevent seasickness (you can always cut a pill in half) or to switch to a brand whose ingredients have less of a drowsy effect. Usually, the physical activity of boating and diving is enough to keep me awake. But suppose that after taking the pill the decision is made to cancel the dive? Some people go home and are zombies for the rest of the day. Not only do they not get to dive, they lose the day for any other useful activity.

The best way to test your tolerance to drowsiness induced by a specific motion sickness drug is to try it out at home. Pick a day when you do not plan to drive or operate heavy equipment, pop a pill in the morning, and see how it affects you throughout the day. If it lays you out like Rip Van Winkle, ask your physician or pharmacist for alternatives when you wake up enough to dial a phone. If it doesn't seem to affect your normal performance, test it on the boat to see if it stops you from throwing up. There probably won't be a perfect panacea.

As a cautionary note, antihistimines might act synergistically with nitrogen narcosis, and exacerbate the effect.

More and more people are getting prescriptions for scopolamine in the form of a transdermal patch that is worn behind the ear like a bandage on a shaving nick. The drug is absorbed through the skin at a continuous rate that is controlled by layers of material within the patch. Not only does the "patch" cause less drowsiness than motion sickness medications with antihistimines, but the delivery of the drug can be discontinued without having to stick your finger down your throat. Although the patch must be emplaced at least four hours ahead of time, if the dive is cancelled you simply peel it off and save it for a rainy day; the drug wears off in a few hours. Furthermore, since the patch is good for three days of continuous use, you can remove it after a single day's diving and have leftovers for another trip at sea.

Patches have side effects, too. Taken in large doses, scopolamine could be disastrous for politicians because historically it has been used as a truth serum; better to puke out their guts than the unpresentable truth. For divers, the side effects are not quite as dreadful: dry mouth or blurred vision. Some people who experience visual disturbances reduce the dosage by cutting the patch in half; they apply one portion and save the other.

When the flags are waving, the divers are heaving.

Some medical practitioners have serious concerns about the use of scopolamine patches for divers. For one thing, the absorption rate is affected by moisture on the skin, meaning that the dosage can vary as a person perspires. For another, scopolamine is a sedative which, by definition, can alter mental functions, impair motor coordination, and cause respiratory depression: effects that continue on the bottom where a diver most needs his faculties, dexterity, and proper breathing rhythm. Although it is not my purpose to scare anyone, you should be aware that scopolamine is one of the belladonna group of drugs and is used to produce amnesia during surgery. It can also induce hallucinations. This may be one rapture of the deep that you can do without.

It goes without saying that the use of any depressant drugs while diving may be inappropriate, moreso for a neophyte whose responses to circumstances are not yet ingrained, and who is therefore unable to react on the basis of experience. Delayed reaction can be harmful.

Get a good night's sleep before a dive. Drink plenty of fluids in order to avoid dehydration. Remember that coffee is a diuretic. And have the common sense to stay away from alcohol.

If the drug scene scares you, try using accupressure. Several companies make wristbands that treat motion sickness by applying pressure to the inside of the wrist, a method that works on the same principle as accupuncture for local anesthesia. For people with suggestible personalities, try a placebo such as ginger, but keep a barf bag handy.

The ailment next to seasickness that is the most bothersome for divers is sinus squeeze. Sinuses are cavities in the head that interconnect the eyes, ears, nose, and throat. They are lined with mucus membranes that are subject to inflammation and swelling, reactions which can clog the passageways and prevent the free flow of air. The common cold is a viral infection that manifests itself with such symptoms as watery eyes, runny nose, sore throat, chills, fever, and, of more relevance to diving activity, congestion of the sinuses. When the sinus passages are blocked, air cannot readily pass between the ears and the nose and throat, thus preventing equilization during descent. This usually means that when you have a severe cold, you can't dive.

But people often dive when cold symptoms are minor and can be alleviated by medication. Decongestants reduce inflammation and reopen the eustachion tubes (which connect the ears with the nasal part of the pharynx), thus permitting divers to clear their ears. I always carry decongestants on the boat in case my head feels a bit "thick" and I think I might have trouble clearing. Pseudoephedrine hydrochloride works for me; this is the generic name for the active ingredient in Sudafed. A variety of decongestants come in the form of tablets, drops, and sprays. Be careful of brands with ingredients added to "improve" decongestion; they may cause adverse reactions at depth.

The major drawback of medically assisted equilization is that the medicament might wear off during the dive; if the sinus membranes swell enough to block the eustachion tubes, the pressurized air that was let into the sinuses during descent becomes trapped; this is called "reverse blockage." The air expands during ascent, causing excruciating pain and possible damage: capillaries may rupture and the tympanic membrane may be torn.

The same thing can occur on the way down, of course, except that then you have the option of halting your descent if you can't clear your ears. There is no way to "unclear" a reverse blockage, and at the end of a dive you have only a limited time before being forced to come up no matter how much it hurts. What this means is that if you use decongestants, you must time the treatments so your dive begins and ends during your medication's most effective phase, so your sinuses will be relieved of unwanted pressure and your head will be clear of blockage. You might call this "planning a head."

Practice clearing your ears naturally on the way down the line. One method is to work your jaw the way you would in a high-altitude jet. Another way is to tilt your head by looking up, then moving the back of your tongue; this is called the frenzel maneuver. Chewing large wads of gum helps develop the proper muscles. Then there is the reliable valsalva technique: hold your nose and blow; but not too hard, because if you have to rely on too much force to clear your ears on the way down, there's a good chance that they won't want to clear on the way back up. If you ever have trouble, either up or down, play yo-yo on the anchor line by moving just a few feet at a time. Do *not* try the valsalva technique during ascent: it will only blow more air into the sinuses, thus exacerbating the problem.

Another ear related problem is vertigo. This is caused by unequal pressure on the middle ears. It sometimes happens to me when I rise too quickly over the hull of a wreck that stands high off the bottom. One ear equalizes but the other one doesn't; the crossed signals result in a whirling disorientation in which the wreck seems to spin in circles and goes out of focus. The dizziness lasts only a few seconds until the other ear clears. I'm used to it now, but the first time it happens to you is likely to be disturbing. Vertigo can also occur during descent; if it does, work your way up the line until the disorientation goes away, then descend again slower.

While we're on the subject of ears, let me mention "swimmer's ear": an infection in which water remaining in the ears after a dive or a long series of dives provides a damp environment that is perfect for culturing bacterial and fungal growths. It can be quite painful. The typical preventive measure for those who are prone to swimmer's ear is to squirt a few drops of white vinegar into each ear after a dive. Some people add vodka or isopropyl alcohol to create a home remedy that hastens drying in the ear through accelerated evaporation. A more expensive version can be purchased from dive shops: the active ingredients in the squeeze bottle are acetic acid (the acid in vinegar) and isopropyl alcohol. The idea is to alter the pH and create an environment inimical to bacteria and fungus. Do *not* pour straight alcohol into your ear in order to get rid of the feeling of water sloshing around, because it can irritate the tympanic membrane and make it vulnerable to injury.

Once you get an ear infection don't try to treat it on your own. See a physician—not your family doctor, but an ear specialist. The diagnosis of the type of infection is difficult to make, and the treatments are different (hydrocortisone for bacterial infections, antimicrobials for fungal infections). The treatment for one could prove harmful if you have the other.

A final word of caution: many people regularly take prophylactic drugs, but little is understood about the effects of pressure on chemicals in the body. Before taking any medication under water, consult a physician who is versed in the medical aspects of diving. If you are so sick that the only thing keeping you going is a massive dose of drugs, then perhaps the best prescription is to lie in bed and read a marine history book while you recover your health. The wrecks will still be there when you're well enough to dive.

"Please don't save me. I can't stand being seasick any longer."
(From *Twenty Thousand Leagues Under the Sea*, circa 1872.)

Fish Bites and Jellyfish Stings

The first question that landlubbers usually pose to a diver is, "Aren't you afraid of sharks?" This annoys me to no end. Not only does the challenge reflect the asker's diffidence, but it makes a presumption of peril that casts diving into the realm of such thrill-seeking activities as parachuting and sports car racing: an image that is undeserved. Danger is what you make of it, and exploring sunken shipwrecks is no more hazardous than hiking through an untamed forest. By contrast, when I go backpacking people never say, "Aren't you afraid of bears and mountain lions?" Instead, they picture the woods as natural flower gardens inhabited by cute and cuddly creatures, they think of mountains as majestic monarchs, they see in rivers serenity and scenic beauty.

Is this disparity an insinuation of man's ancestral obeisance to the great outdoors, where he lived and evolved, coupled with an innate fear of the sea? Or is it just bad press? Philosophical digressions aside, a cursory glance at today's communication media reveals intentional slants. Books and magazines geared for the non-diving masses peddle danger because it sells copy, while diver-related publications stress fun and enjoyment because they advertise vacation packages. Similarly, Hollywood movies make fortunes on death and mayhem while public broadcasting stations glamorize the ocean's natural wonders. Between these divergent viewpoints there is a happy medium.

The truth is that people get bit by sharks. The truth is also that you have a better chance of being struck by lightning.

Generally speaking, if you don't mess with the sharks, the sharks won't mess with you. Unpredictable predators that they are, a fully rigged diver making noise and blowing bubbles is a pretty imposing figure to the cartilaginous species. In most of my experiences with sharks, they were so skittish that I couldn't get close enough to take pictures. In a few instances they displayed curiosity and brushed close by, until I reached out to touch them and scared them off. Once a shark bit my strobe—but that was a function of the animal's sensitivity to the electrical impulses of the recycling capacitor; they are disturbed by the inductance of electrical current.

In short, be wary of sharks but do not harbor unreasoning fear. Sightings of sharks under water are so rare that people usually race for their cameras when the word gets out that one has been spotted. There is nothing more ludicrous than a boat full of divers chasing after a shark like a pack of hound dogs running down a rabbit. You can get away with such behavior with most sharks, but don't try it with a grizzly or puma.

Yet, there is no denying that the underwater realm provides a habitat for animals which, while seldom deadly, can cause pain and injury if approached or mishandled. Books are available that describe the death throes induced by the bites of sea snakes and the sting of cone shells. Neither these nor any other

Jellyfish and goosefish: no terrors of the deep, but they can cause pain.

aquatic animals prey upon man for food, nor do they go out of their way to attack. Sea snakes are territorial, and bite only when their personal spaces are invaded. Cone shells must be handled in order for them to inject their venom. The same is true for other poisonous animals such as scorpion fish, stonefish, lionfish, stingrays, and electric eels. If you leave them alone, they'll leave you alone.

Most injuries occur accidentally, such as when you kneel on or lean against a creature whose protective coloration camouflages it from the surrounding bottom features; or when you pass by or stick your hand into a hole inhabited by a moray or lobster; or when a sea wasp or jellyfish drifts into you. But while these bites and stings may hurt, the pain is usually short-lived and permanent damage almost never results. The most serious complications arise from infection and allergic reaction.

There are several preventive measures you can take against marine life hazards. Begin by acquainting yourself with the local wildlife—under water as well as on the streets. Read up on endemic fauna and learn to identify those creatures best to avoid. Ask the boat crew and dive shop personnel for advice on the prevalent harmful species, and for precautionary suggestions; no one knows the regional waters better than those who dive in them on a regular basis.

Don't grab such well-known nasties as goosefish, wolffish, sculpins, barracuda, or anything with a mouth full of ugly-looking teeth. My friend Mal Kroeber was once bitten so bad on the head by a triggerfish that his scalp required stitches. (Four of us got bit by triggerfish that day—our best guess was that they were spawning on the wreck.) Stay away from squid and octopi unless you want a little bit of the beak. Puncture wounds can sever nerves—a far more serious injury than the mere tearing of flesh. And wear protection.

Rubber can be a prophylactic against many kinds of ailment. Under water, the same neoprene that protects you from the cold also shields you against such annoying and painful abuses as sea urchin spines, bristleworm stings, pointed barnacle shells, and fire coral (which is actually a hydroid). Many times on the anchor line I have felt the sharp stings of jellyfish or invisible plankton across my cheeks and upper lip; I was glad I was otherwise fully covered. There have been occasions when I had to turn my face away from the current because there was no way to avoid getting stung by the drifting hordes.

My worst experience, however, was with a Portugese man-of-war. I was covered from head to foot but, because the water was warm, I opted not to wear gloves. I noticed nothing unusual when I jumped off the back of the boat. What I could not see was the distinctive purple balloon floating alongside the starboard hull. As I swam toward the bow, the current forced me up against the boat, so I pushed myself away—and put my hand into the long tentacles of the most dreaded coelenterate of them all. I jerked back as if someone had poured acid on my hand; the burning was intense. Within seconds I was back at the ladder, but already my fingers had swollen so much that I could not bend them at all: they stuck straight out like fat sausages.

I climbed aboard by hooking my elbow around the upright of the ladder. Two crew members helped me to the bench, then had to get me out of my gear. The tentacles were so deeply imbedded in my skin that the mate had to scrape them off with a knife. Then, for two hours I lay on the deck groaning in agony; all the time my hand felt as if a blow torch were being played across it. My forearm swelled, and the muscles felt as if they were being twisted through a ringer. Tingles crept up my arm and halfway across my chest before the spread was arrested.

The excruciating pain went on and on. That night I was still ill and aching. Yet, I was lucky. I had come in contact with only a small portion of the animal, I had gotten only a mild dose of poison, and I was not exceptionally

Wolffish and trigger fish: teeth to be avoided.

allergic to the toxin. In severe cases a person can suffer respiratory paralysis. No breath means death.

Did my run-in with a Portugese man-of-war stop me from diving? Yes—for the rest of that day and all of the next. Did I profit by the experience? Absolutely—I always wear gloves and I look hard before I leap. I also accept the fact that life has more surprises in store for me.

My most ignominious diving accident came not from a barracuda but from a lowly bergall. The four-inch fish darted in between my mask and regulator and took a chunk of meat out of my upper lip. My lip swelled as if I had been punched in the mouth, and it was weeks before the wound scabbed and healed. All the time, I had to explain to my friends how I had been attacked by a voracious sea monster the size of my palm.

I don't mean to pooh-pooh dangerous marine animals, only to put them into perspective with the real hazards of the underwater world. With the alarming increase of ciguatera, you are more at risk from eating fish than from being eaten by one. In that case, the voracious appetite more likely to result in personal injury is yours rather than a shark's. Get rid of your bang stick *and* your fork, and you'll be much better off.

Green moray and sand tiger shark: over-maligned predators more interested in prey smaller than man.

Survival at Sea

It was recently brought to my attention that I am acknowledged as the expert in getting lost at sea: a rather dubious honor I have earned not because of my lack of navigational skills, but because I have gone adrift and have been left on wrecks more often than anyone else in the business. The reason for this, I hope, is merely a matter of chance: if you dive long enough, often enough, and under conditions severe enough, it is bound to happen eventually. Therefore, I plead innocent to charges and claim to be a victim of circumstance. Divers get lost for a variety of reasons, most of which are out of his control. So, if there is a moral to this chapter it is the boyscout motto: be prepared.

Two scenarios: in both I picked my way through turbid water till I returned to the location of the anchor line. I could not find the grapnel where it was supposed to be hooked into the wreck. Naturally, I figured I was disoriented, and began a series of sweeps to see if other parts of the broken down, wooden hull looked more familiar. They did not. I spent a few minutes searching, but not too many. It's important to know when to keep looking and when to give up and make alternative arrangements for getting back to the boat. If you spend too much time swimming around in circles you're likely to run down your air supply too low for safety, and you might exceed your no-decompression limit. So, what do you do?

If I say, "Come up," you might think that's too obvious a solution. Yet, I have known divers who were so dependent on the anchor line, who had such an ingrained mind set about ascending it as the only way to get off the bottom, that the inability to locate it threw them into a fright; they continued to hunt frantically until they were in serious trouble, until they nearly ran out of air and were forced to scream for the surface, where they arrived shouting for help and unable to breathe or think about inflating their BC's or switch to their snorkels. This is not an appropriate response to the situation. So the question this reaction begs is, "Why didn't they just come up sooner?"

The answer: because they lacked the confidence gained through experience to go back to the boat on their own, without the anchor line feeding them reference information. This kind of insecurity, this perceived threat of danger, is not all that uncommon; but usually those people don't make it past pool practice. A person with such severe uncertainty about his capacity to abide the seven seas without being connected to the boat by a continuous static line, or with an irrational fear of open spaces (agoraphobia), is an unlikely candidate for ocean diving where horizons are not visible from eye level. You need to understand that the anchor line is not the *only* way back to the boat, it's just the most convenient way. You can get back without it.

On the other hand, there are times when a misplaced anchor line has a more serious implication. If you have definitely ascertained that the grapnel is gone, there's no sense in looking any further. It goes without saying that if the grapnel is no longer hooked into the wreck, neither is the boat that's attached to the end of it. Which means that when you reach the surface you might find yourself alone among the whitecaps. Now *that* is something to panic about. Then, after you get over the panic, change gears into survival mode and begin taking strokes to ameliorate your situation.

Being lost at sea is no fun; the recollection still makes me shudder, even in the security of my study. My own experiences apart, there are an unsettling number of documented cases in which divers have been left on wrecks without transportation back to shore. The most common cause is a divemaster who miscalculates the head count, tells the captain everyone is up, then goes down below and breaks open a six-pack for the long ride home. In one incident I remember, the diver was not missed until the boat docked and the captain started collecting money; when the till came up short he noticed all the extra dive gear left aboard after everyone else had unloaded. How embarrassing.

Although this situation is more likely to occur on "cattle boats" (big charter boats carrying a large number of "walk ons") it can happen on any boat on which people are strangers to each other, and in particular if an individual is not known to the group at large. When people look around and see familiar faces, they're not likely to remember the new kid who walked on that morning and sat quietly in a corner all the way out to the wreck site. Make yourself known, introduce yourself to the captain, crew, and fellow divers: not just so they'll remember you for the countdown, but so you can strike up new friendships and join in the camaraderie. And make sure your name is on the divemaster's list before you enter the water.

Human error is not always at fault. A diver can come up too far away from the boat to be seen or heard (if no one on board is paying attention), or he can come up in the dark or in a fog (weather or mental). In another incident I remember, a diver missed the boat because a strong current carried him away, and those on board could not hear him shouting because a compressor was filling tanks on the after deck. He was soon swallowed up in a pea-soup fog. When his buddy showed up and told us they had become separated on the way up the line, we figured he must have passed out and fallen back down to the bottom, unconscious, and died. Nevertheless, two people went out in an inflatable chase boat to look for him just in case he was adrift (best case scenario) and the Coast Guard was called to send out a helicopter for aerial reconnaissance. The Coast Guard responded quickly, but airborne observation was hampered by the fog. The chase boat carried a compass, radio, and radar reflector (a hastily contrived "sail" of aluminum foil made from sandwich wrap), and was fortunate enough to find the diver about two hours later. When they radioed the good news, in true gallows humor we took bets on how much gear he still had with him. He had dropped everything except his light and videocamera, and he was just about to let go of the videocamera.

There have also been cases (not in this country) in which the dive boat left the scene with divers in the water—and the bodies were never recovered. I also know of one case in which the dive boat was swamped by an errant wave; it sank on a shallow Florida reef while most of the divers were on the bottom. Admittedly, it was a small, center-console job, but nevertheless, it can happen. Dive boats have been hit by traveling freighters and, while I do not know of any that were sunk by the collision, some were severely damaged and divers were thrown overboard by the force of the crash. All of these happenings can place a diver on his own, with survival his primary concern.

So why didn't they tell you these things when you took your certification course? After all, when you take "driver ed" in highschool they show you films of highway accidents and horribly mangled bodies. Why not here? Perhaps, I will allow, such information goes beyond "basic" scuba; perhaps, they figured, you will pick it up later. On the other hand, perhaps, they just didn't want to frighten you with horror stories that might make you quit the course. After all, ninety-nine percent of diving is safe and simple fun. The same is true for driving, but we wear seatbelts anyway.

Admittedly, the incidents just related are isolated events—but they happened, and they will continue to happen. If you decide to pursue wreck diving assiduously, you should understand *all* the risks, not just those that are most likely to occur. And, as this book has already illustrated, wreck diving may carry you to some far out places, far offshore, and far away from immediate aid, so that some insight into the measures available for self-rescue and help-in-rescue can put you more at ease and increase your chances for survival at sea. Keep this in mind when I say that I *want* to scare you—at least enough to get your attention.

So let's go back to the two scenarios I introduced at the beginning of the chapter. Once I determined that the grapnel was gone or that I was lost, I got out of there while I still had a broad margin of safety. In the first case I tied my wreck reel to a solid timber, made sure the line was not chafing and was clear of sharp objects (barnacles and, on a steel wreck, rusted metal edges), got positively buoyant, and made a slow ascent while unreeling the line and monitoring my depth gauge and timepiece. The reason for this attachment to the bottom is that in the featureless immensity of the sea a shipwreck is a known reference point. After reaching the surface and doing a couple of quick three-sixties, I discerned that there was no boat in sight.

Not to worry. I knew the captain would come after me, and I knew that the first place he would look was on the loran coordinates for the wreck. I waited patiently, and ten or fifteen minutes later I saw the white painted bow splashing through the waves. The reason he was gone for so long was that other divers were on the line when the grapnel pulled free, and he had to keep the engine disengaged from the propeller until they all clambered aboard. I was not mollified when he passed by and said as he leaned over the rail, "It's a good thing for you I had to come back for my buoy," which he then proceeded to pick up. Of course, I knew all along that he would come back for me: I hadn't paid for the charter and my new Nike's didn't fit his feet.

In the other circumstance I ascended without tying off to the wreck because surface conditions precluded the successful execution of such a maneuver: a strong current was running and high waves were crashing; I would not have been able to hold onto the line. The problem with this situation is that a diver is swept away from the site with the current, while the boat is more at the mercy of the wind. Since wind direction and current set do not usually coincide, the result is a divergent path in which the boat and the diver are impelled along different routes. Once the captain is able to power the boat back to the wreck, he must then run a compass course down the heading of the current. Again, I was found without too much difficulty, mostly because I was accompanied by an inflated liftbag.

On a couple of other occasions which I won't bother to relate, I was adrift for as long as an hour. That doesn't sound like a long time: it's only the length of two sitcoms, a more unendurable torture. But being alone at sea is not the same as sitting in front of the glass teat. You *know* when the sitcoms are going to end, but you can never be sure when, or if, your rescue will be effected. It's that uncertainty lingering at the back of your mind that is so agonizing.

The most important aspect of staying alive in life-threatening situations is proper attitude. It is unfortunately true that some people give up mentally, and die, before their bodies reach the extreme of physical endurance: the result of a weak temperament which is either a low tolerance to pain or stress, or a feeble will. There is very little that can be done for an individual so predisposed—but then, that kind of person will not usually pursue an activity that immerses his body in a potentially harmful medium: he will either bowl or play golf, or confine himself to Saturday night bridge parties. By implication, one who ventures knowingly underwater has an edge over the landlocked faint-hearted: an innate self-confidence that will stand him in good stead when the chips are down. Furthermore, that inner strength can be made stronger by conscious development—either from within (as you gain assurance of your physical abilities through experience) or from without (by taking advanced training courses). The more familiar you are with your own capabilities, the better you will be able to handle yourself in a tight spot.

Next in importance is to keep firmly in mind the support structure offered by society. No person stands truly alone, nor even close, in a world with such high regard for human life. I have found it positively enthralling to be on a boat during an emergency, when I could behold the deep concern displayed over an injured diver, or when I could observe the coordinated effort to help someone in distress. People *care* about their fellow man, and will do everything in their power to help a person in trouble. And I don't mean just those people in positions of responsibility, like dive masters and boat captains, but everyone. When the going gets tough, wreck-divers shine their brightest.

Then there is the formal support structure provided by the government in the form of the Coast Guard: in my opinion the noblest organization of professionals in the history of the world. Although the Coast Guard is primarily concerned with drug interdiction and vessel inspections, the conduct of its search and rescue operations is second to none. To a diver in distress, the sound of an approaching helicopter is the modern day equivalent of the bugle call signaling the arrival of the cavalry: whirling blades instead of blazing saddles. For large-scale situations the Coast Guard can also dispatch boats and planes, all of whose personnel are fully trained, highly motivated, and thoroughly proficient. It's a comfort knowing they're there if you need them.

The U.S. Coast Guard: a diver's best friend in time of need.

What all this means is that should you find yourself drifting off into the sunset, you can rest assured that everything humanly possible will be done to effect your rescue. That alone should put your mind at ease when considering your plight. I do not mean to imply, however, that you should lie back on your inflated BC and relax until you hear the clarion call: it may be angels coming to take you away instead of the Coast Guard.

My advice is to be prepared for the worst. Translated into wreck-diving language, this means that mental preparation and equipment purchase can save you a lot of grief, and furnish you the wherewithal to stay alive and aid in your location. Let's talk about gear.

I've already mentioned the wreck reel as an essential item that has several uses. Add a liftbag, and instead of using your own buoyancy to maintain station over the wreck, you can let the liftbag do the work while you merely float alongside. If you go adrift, an inflated liftbag will act as a marker and will be more highly visible than the basic black exposure suit. Orange liftbags work best.

The simplest article you can add to your equipment list is a dime store whistle. Cheap plastic will suffice, although I've broken a few due to the rigors of gear stowage. A brass whistle will last until you lose it. I tie my whistle onto my regulator hose a few inches up from the mouthpiece, where it's out of the way but always available. The piercing shriek can be heard over a much longer distance than a frantic shout, even better than a woman's high-pitched scream, and can cut through the throb of running machinery and reach people chatting merrily in the cabin. It's worth having just to call for assistance in climbing up the ladder when no one is around.

A step up from the whistle is the "acoustical signaling device," a fancy name for an airhorn that connects to your BC inflator hose and uses low-pressure air to curdle your eardrums and those of everyone within the blast radius. When you hold the ASD out of the water and press the actuating button, make sure to turn your head or cover the adjacent ear with your other hand. The noise is quite loud.

Now let's go from auditory warnings to visual aids. Whatever you can do to make yourself more visible is appropriate. Brightly hued exposure suit material is desirable under all circumstances except clandestine military operations. And I don't mean color coordinated neoprene with matching designer stripes. Plain orange or red will suffice, especially on the head, sleeves, and upper torso.

An inexpensive add-on is retro-reflective tape. Unlike reflective tape, which only reflects light that strikes its surface at a 90° angle, the crosshatch pattern of retro-reflective tape reflects light from nearly all angles. Two intersecting strips laid over the top of your hood will reflect aerial searchlight beams with a clear message: X marks the spot. Additional strips across your shoulders or along your sleeves will increase the size of your X, an idea you can reflect upon while floating around at night. Your liftbag might look better with retro-reflective tape, too. Retro-reflective tape comes in rolls, has a peel-off backing, is self-sticking, and cheap. However, I have found that the glue does not adhere to neoprene very well; the tape soon peels off. Attach the tape with an adhesive such as epoxy or tube cement.

Another bargain buy is the signal mirror, largely for daytime use when the sky is not overcast. It comes with a sight hole for aiming. You look through the hole at an approaching rescue craft, then tilt the mirror back and forth, allowing it to bounce the sun's beam when the angle is right. Get the plastic model because it's unbreakable unless you let your kids play with it. A wrist lanyard attached through the loop hole will prevent you from dropping it and ruining your whole afternoon. Some models also have one side coated red for night signaling—no, not using the full moon as a light source, but your dive light. In this case it is advisable to alternate sides because Coast Guard pilots wear night vision goggles that block wavelengths at the red end of the spectrum.

More costly but definitely worth the investment is a waterproof signal strobe (not a camera flash). The intense burst of light is visible for miles even

Left: Retro-reflective tape sewn to a hood. Right: A small dive light housing (with the batteries removed) makes a dandy waterproof survival kit.

in daytime, and is not dependent upon local weather or environmental conditions. The repeating beacon is like a miniature lighthouse, not only giving away your position but emitting a continuous update on your whereabouts. The Coast Guard rates the strobe as the number one detection aid; its rescue swimmers do not go in the water without one. Remember, though, that if you take down a strobe to mark the anchor line, you need another one in reserve for emergency drift signaling.

Photographers can flash their camera strobes manually when they detect the approach of rescue craft. Not only will the intense light give away your position, but if you actuate the strobe by depressing the shutter release button you may get a great photograph of the rescue operation.

One way to increase your vertical profile is with a "safety sausage." This is an inflatable tube that starts out the size of your fist, but can be orally inflated to a length of about ten feet. It is so low-priced that it can be considered a throwaway item.

Sometimes useful is dye marker. Life rafts and pilot bailout kits have long been supplied with packets of powder which, when dumped into the water, create an emerald green stain that contrasts sharply with the dark tint of the sea. Dye marker is usually deployed after the rescue craft come within visual range. However, in swift current and high waves the chemical disperses quickly.

A chemical light may have some value at night or on dark overcast days. It is a slender plastic wand that is snapped to break the barrier between two

Left: A flare gun kit that works when wet. Right: Ed Dady sends smoke signals after the cannister was taken on a dive and exposed to the water.

144

 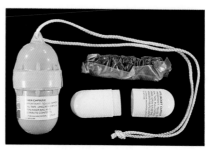

Left: A signal mirror and plastic whistle are cheap insurance. Right: This dye marker packet comes in a waterproof casing.

chemicals which react together to create light. The resultant glow is dim, but long-lasting. The same caution that applies to the red signal mirror must be noted here; the green sticks may not be visible to pilots wearing night vision goggles. In any case, I think the value of a chemical light is dubious at best: kind of like a life insurance policy that pays off only if you're struck by ball lightning at midnight on a sunny day in a month with seven letters.

The big brother of the chemical light is the emergency flare. These are difficult to obtain because of strict transportation regulations, making it a headache for marine outlets to stock; some don't even bother. When available, though, you should pick up a couple for that especially remote dive location. They can be quite effective in attracting attention from a distance. The rocket-propelled parachute flare will reach an altitude of over a thousand feet. The hand-held variety is less expensive, but doesn't climb as high.

The best kind I've used is the military type, about six inches long and two inches across, with a high intensity flare on one end and a smoke bomb on the other: flare for night use, smoke for day. Either end is activated by pulling a cord. Hold it high over your head for best visibility, and don't worry about inhaling the smoke: it's non-toxic. Lately, the only place I've seen them for sale is in military surplus stores, and even there they're becoming rare.

Now let's assume that you've gone adrift, with or without any of the specialty equipment I've suggested. What should you do? First of all remain calm, evaluate your situation, and make deliberate decisions. Then, ready yourself for a long siege. Sure, you may get picked up before you have time to

Barb Lander tests dye marker in the ocean. Notice the size of the dispersal pattern after about five minutes.

catch your breath—but, then again, you may be out there all night. Always hope for the best and prepare for the worst. Husband your strength and your other resources, including the air in your tanks. Do *not* discard any of your equipment, not because you paid a lot of money for it, but because you never know what you might need.

Inflate your BC or drysuit. Make yourself comfortable. Relax. Do not struggle. Go with the flow. Look for the boat but don't get alarmed if you can't see it. Assess your physical condition—are you hot, cold, wet, or hurt? Do your best to ameliorate any discomfort, but have the will to suffer short-term pain and emotional stress. Intellectually, you know it won't last forever.

I once read a statement made by the survivor of a torpedoed merchant ship. He said that when he was finally rescued, after more than twelve hours in the water, his worst injury was his neck—from constantly turning his head to look for passing vessels. I can sympathize with him. It is difficult to let a minute go by without scanning the horizon for signs of rescue craft. Your best bet is to face up-current: the direction from which your own boat is most likely to come.

Do not lose your mask; it will keep water out of your eyes and the rim will act as a shade. If you want to remove it for nasal breathing, pull it down over your neck. In rough seas you'll find it better to keep your mask in place and breathe through your snorkel, you can rest lying face down. If you want to maintain a vertical attitude, position the snorkel so it points up off the top of your head—a snorkel with a noncorrugated flexible hose connecting the mouth-piece with the riser tube is best. Try not to swallow sea water: it may make you sick, and vomiting will dehydrate you.

Implement surface signaling procedures as appropriate, but don't waste your strength or expend emergency equipment needlessly. Conserve the batteries in your lights and strobe. Constantly re-evaluate your situation, and be prepared to change tactics should the need arise.

In all likelihood you will suffer very little physical discomfort other than heat or cold, and possibly having to urinate. Your biggest problem will be dealing with the psychological aspects of your plight. This is where your inner resolve will stand you in good stead, will see you through the unfortunate predicament. Put aside irrational thoughts, and don't think about sharks. (You wouldn't think about lions or tigers or bears if you were lost in the woods—unless it happened to be the Emerald Forest.) Just do what is necessary to stay alive. And begin thinking about the day you're going to look back on this experience and laugh about it.

Remember the words of Winston Churchill: "Never give in. Never, never, never, never give in." Help is on the way.

Room for Ladies

Women in diving is a hot topic of discussion these days.

More and more women are entering fields of endeavor that have previously been dominated by men. This is causing not a little concern, and is creating disaffection between the sexes when, biologically if not culturally, men and women have been predisposed by evolution to share the activities of life. To be sure, mankind's history of division of labor has played an important role in the survival of the species. When people lived in caves and were preyed upon by saber-toothed tigers, it made sense that those individuals who had the physical attributes to give birth and suckle the young should dwell in a position of relative safety in order to protect their charges, while those who were otherwise dispensable to childbearing and child caring should be expelled into the wild in order to perform food-gathering functions with equivalent longterm survival value.

Today we are in the midst of a cultural revolution which seeks emancipation from the mores of the past, and which aspires to supplant atavistic ice-age attitudes with mature doctrines of life more consistent with the genuine instabilities of modern civilization. Instead of tigers we have terrorists, instead of hunting we have shopping, and instead of the constant threat of bodily harm we have mortgages to pay, jobs to keep, and families to support: emotional pressures rather than physical perils, but every bit as threatening to our continued well-being. Unfortunately, while the saber-toothed tiger has become extinct, and caves are no longer our preferred place of abode, Neanderthal man can still be found haunting the side streets of our splendid cities and neighborhoods.

This is not to say categorically that chivalry has been replaced by chauvinism, that inequity and inequality run rampant, that opportunities are denied to women, or that men breed with contempt. But to ignore that double standards exist also ignores reality.

Confusion reigns when sexual dimorphism is misunderstood or its significance is misapplied. By definition, ''dimorphism'' means two distinct forms within the same species. "Sexual dimorphism" means that the distinct forms are manifested by anatomical differences between the sexes. Sexual dimorphism is common to nearly all forms of life on Earth, both animal and vegetable, that are higher on the evolutionary scale than yeast and the amoeba, and is the foundation for gene mixing and heredity variation via sexual reproduction.

Problems arise when people interpret "differences" to mean "inferiorities" or "inadequacies" or "weaknesses." Science makes no such inference, nor has any such correlation been established. Sexual dimorphism is a biological principal that does not require sublimation: it is a fact. If you cannot accept that, stand nude in front of a mirror sometime with a member of the opposite sex. Even to a solipsist, seeing is believing.

In the grand scheme of nature, specialization is as important to the survival of life as speciation. Similarly, the balance of modern society is maintained by people specializing in various occupations: some people build houses, some people sell houses, some people repair houses, and some people live in houses while performing other jobs. Biological necessity has been replaced by cultural ethic. Indeed, is has been suggested that man's biological evolution ended when he developed conceptual intelligence and gained control over his environment and, by extension, his destiny, and that further evolution will not be physical, but cultural.

So what does any this have to do with wreck diving? More than I ever imagined or wished to learn, until I interviewed several female divers whose opinions I respect, and discovered that iconoclasm has a long life to live. You see, all I wanted to know was how women adapted to equipment that was designed for the male physique. What I got was a diatribe on gender bias, machismo, caveman mentality, and that popular phrase ever more in the news today, sexual harassment. (And those are just the polite examples.) As I pulled the nails from my hands and feet, I determined rather quickly that women in diving have more to overcome than a tank that is too tall for the female torso.

The question this begs is: what is to be done about it? We could shrug it off and say, "Boys will be boys, and so will men." Maturity comes late in some people, never in others. But, although during the interviews I endured some uncomfortable moments as the dartboard of misplaced aggression, I

A pair of crayfish demonstrate sexual dimorphism and the eternal tussle for position. Mating is healthier than berating.

The ideal dive team is one that is bonded by love and sharing. Baby sitters take care of the kids while Molly and Alan Troutman dive together.

cannot help but wonder why, if people in today's society have the freedom and opportunity to choose the kind of person they want to be, so many have made such a poor choice. After all, proper behavior is an attribute that is not only instilled by parental guidance, it can be molded by a conscious decision on the part of the individual. That is what we can do for ourselves. But what can we do about the behavior of others?

Quite simply, don't accept it.

I know this must sound like a cop-out, but if *you* are not willing to stand up for your rights, why should you expect anyone else to stand up for you? You can object passively by refusing to dive with those who continue to carry the caveman routine beyond its useful function; dive elsewhere. Or you can protest actively by calling attention to unacceptable behavior; it's amazing how effective this is, seemingly because those who are used to intimidating others don't know how to respond when you bounce it back at them. Perhaps you can't teach an old dog new tricks, but people, if they get enough negative feedback, will learn what they cannot get away with. Of course, you'll be branded as a conceited pain in the beaver tail, but what goes for body building also goes for mind setting: no pain, no gain.

This somewhat nebulous discourse was not intended to be so long-winded or deeply philosophical, but perhaps it was necessary in order to warn women what they might be up against and to enlighten men about what they should be sensitive to. Besides, my interviews opened such an explosive can of worms that I felt I would be neglecting my duty if I did not share those worms with my readers.

Of course, every story has two sides . . .

No matter what I say or how delicate I turn a phrase, there's no way I can escape this subject unscathed. So why not delete the chapter altogether and round off the book at an even dozen, instead of tempting superstition with a bad luck number? It would certainly be safer. The reason is that I enjoy attacking accepted doctrine when that doctrine is irrational and in need of change. Heresy has a check-and-balance effect that forces society to re-examine its values, the same as introspection does for the individual. By drawing the controversial sword I permit my female counterparts to provide a foil; the inevitable debate and exchange of ideas will help stimulate a more thorough understanding of the human psyche, social interaction, and the problems of personal relationships. Besides, I can offend only those women who are either unreasonable or emotionally insecure: a decidedly infinitesimal number.

In addition to the undeniable physical differences between men and women, there are psychological differences as well. Why this is so is not understood, and, anyway, is a matter far beyond the scope of this book. Nevertheless, men and women respond differently to situations of stress, behave differently in consequence of their actions, and approach problems differently. (Read "differently" as "differently.") Also, "men" and "women" are statistical stereotypes and are not intended to apply to any individual.

I cite an illustration that was explained to me by one of my female advisors. Women (stereotypically) do not perform in response to the "macho factor" that motivates so much competition among men. Whereas a man might dive heedless of conditions and of his ability to perform under those conditions—due either to male-oriented peer pressure or to a need to show off for the "girls"—a woman is more likely to do what is best for her under the circumstances, and her female companions are more likely to accept her decision without scoffing.

Trouble begins when a capable woman dives under adverse conditions and less capable men then feel compelled to follow her example or else lose face. This the result of a culturally induced ideal that declares men as bolder and stronger than women. So when a woman unintentionally competes with a man where it hurts him the most, the man might retaliate in a variety of ways: everything from the cold-shoulder routine to open hostility.

Yet, diving is an activity whose performance is governed by skill, experience, familiarity with equipment, and that undefinable quality known as talent. It has little to do with brute strength. Nevertheless, the histrionic hype that pretends to associate diving with danger, daring, and defiance of the odds (stereotypic male-domineering perceptions) enforces the belief that men make better divers than women. If there are more male diving experts in the field, it is largely a matter of priorities and self-fulfilling prophesies. Perhaps women have better things to do with their spare time, or perhaps they don't have as much leisure.

Physical stamina counts more than strength. Despite their smaller proportions, women who are firm and fit are in better diving trim than men who are overweight and out of shape. In part this is due to the aerobic nature of diving, in which a conditioned cardiovascular system is more significant than sinew in propelling a person through the water. In addition, the buoyancy provided by the water and by compensation through mechanical means neutralizes the weight of the gear, placing men and women on equal "finning."

What is more important than pure mass of muscle is muscle to mass ratio: that is, how strong a person is in relation to body weight. A three hundred pounder can lift a set of doubles as if it were an empty suitcase, but probably can't do a single chin-up.

This is not to say that strength is not important, such as for climbing boat ladders. With shorter legs and smaller thighs, negotiating narrow rungs while wearing heavy scuba gear can present a very real problem. Climb up the ladder on your knees, take off your gloves for a better grip, and ask for help if you need it. On the other hand, if you play the male/female role playing game, and try to manipulate men into loading your gear on and off the boat, not only will you earn no respect, you'll gain a reputation as a dive boat bimbo, and deservedly so. Better to buy a collapsable luggage carrier for those heavy tanks, or carry them on your back by slipping into your harness. Be sure you understand the difference between asking for assistance and responding in kind, and getting someone to wait on you.

Learn to be self-sufficient; adopt the attitude that you can succeed on your own without male attention. Be proficient with your equipment; lack of familiarity with gear on the surface implies a similar deficiency underwater, and will incur derision. Engage in wreck diving not as a female who demands special consideration, but as a generic and genderless wreck diver who has as much to learn as an upcoming male. But don't be content to simply tag along after male divers and let them make all the decisions; take a leadership role. Communicate with successful women divers and learn how they overcame the difficulties of apprenticeship. And keep in mind the motto of the all-women mountain climbing expedition to K2: "A woman's place is on top." In diving, perhaps her place is on the bottom.

You can do all this without having to compromise your femininity. You may find yourself despised by some men, but you'll be admired and appreciated by the ones who count. And should you ever doubt your ability to match men in the water, remember that wreck diving is a matter of brain over brawn. A woman may have a mind of her own, but not only is that mind at least the equal of the minds of men, it is also unhampered by the hang-ups of dominance and the handicap of machismo. It's a wonderful little machine when used to its full potential capacity.

Aside from the justifiable concerns of color coordinating their equipment, what other gear problems do women have? Since I've already mentioned tanks and torsos, I'll start from there. Many women complain that scuba cylinders are too long: both a headache and a pain in the rear. The obvious solution is to get shorter tanks, say, ones that hold 50 cubic feet of air instead of 80. Your initial reaction may be to protest that you'll get shorted on bottom time. Not necessarily. Is there any truth to the claim that women are better on air then men? Absolutely. Generally speaking, a smaller person has a smaller tidal volume and less body mass that needs oxygenation; she breathes less air. That's why you don't see weight lifters winning the Boston marathon.

If 50 cubic feet is not enough air, get doubles. Twin tanks have an advantage that is often overlooked: better balance. A single tank rides on the crest of the backbone with considerable posterior extension. It has a tendency to lever you from side to side like a tail wagging a dog. Add a pony bottle, and an already inherently unstable system loses even more equilibrium. But twin 50's straddle the spinal column and hug the back with comfort. Furthermore, climbing boat ladders is easier in doubles because of the reduction in backward leverage. You don't have to get a manifold or crossover bar; you can go with twin singles and separate regulators, letting each tank and regulator act as a backup for the other. Many women who wear twin tanks use this arrangement instead of doubles with a pony.

That dive gear is mass manufactured for the mass market may seem too obvious a concept, unless you understand that women are not the major purchasers. (Or, at least, they are not the major end-users.) Because diving is a male-dominated activity, equipment manufacturers cater to the needs of those who use their products the most. No prejudice is intended. It's just plain good business sense in a tough economic climate when sales are low and profit margins are slim. What this means to women is that some dive gear doesn't fit right.

It's simple enough to replace a regulator mouthpiece that is too large for your mouth, but when it comes to thermal protection a woman might find herself left out in the cold. Rental wetsuits can be a problem. Not all dive shops have the volume of business to stock every size. This forces compromises. A man's wetsuit worn by a woman of the equivalent height and weight is way out of proportion: the hips are too tight, the waist and chest too loose; water will flow freely around the core area. In a woman's two-piece, the bottoms may fit but not the top, or vice versa. Convince the staff assistant to split up the suits so you can rent a bottom from one set and a top from another, so you get a match that fits. A better solution is to buy a tailored wetsuit.

Some drysuit manufacturers seem to think that flashy colors and bright dazzle stripes turn a man's drysuit into a woman's. If so, they are underestimating—perhaps even insulting—the female intelligence.

On a photo shoot with Joyce Hayward, we took turns posing for each other's cameras.

A drysuit that is too tight around the hips violates the thermal qualities of the underwear by pressing it flat; it's the fluff that retains heat. There's room for busts of all sizes, but sleeve lengths are rarely short enough and booties are almost always too large. This means costly alterations.

Pay special attention to long hair. Worn loose without a hood it gets caught in tank valves or the mask strap; braid it or put it in a ponytail. Keep a bottle of leave-in conditioner in your gear bag; after the dive, spray on the conditioner so you can get a comb or brush through your hair. For drysuit diving, wear a shower cap or latex swimming cap to help get your head through the neck seal. Gene Peterson cut a round hole in the nape of his wife's drysuit hood so her ponytail could be pulled through, instead of leaving it bunched up around her neck. Don't use rubber bands on ponytails: they have a tendency to snarl wet hair; instead, use wide band ponytail holders made of cloth.

Neoprene neck seals are more prone to catch long hair than latex. First, because latex is more stretchable, less strength is required to get your head and all that hair through the seal. Second, latex does not roll up in your hair like neoprene will. If you have a neoprene neck seal and it causes you grief, have it replaced with latex.

Remember, too, that long hair stays wet longer than short hair, is not as easily dried, and will cause more significant chilling as the extra quantity of water evaporates. Bring an extra towel for your hair, and a warm hat to wear after the dive.

Lead weights can pose problems for women. Worn around a slender

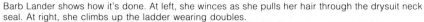

Barb Lander shows how it's done. At left, she winces as she pulls her hair through the drysuit neck seal. At right, she climbs up the ladder wearing doubles.

waist, there may not be enough room for the buckle in front, or the strap may bend so sharply over the weight adjacent to the closure that the pressure keeps flipping open the locking lever. Try stacking weights: instead of threading the belt through one after the other, put two or three atop each other and make one loop through the entire stack.

Molded, curved hip weights are more suitable for men with an A-frame body. For women with a large hip structure and a waspish waist they may cause discomfort because of the way they rest on the hip bone, and because the angle of curvature is not contoured for the female anatomy. Furthermore, improperly placed weights can pinch the femoral arteries and cause tingles and numbness; these warning signs indicate restricted circulation that increases the risk of DCS.

Women bruise easier than men. Too much weight contacting the hip bone can cause contusions that will become painful after a while, especially if you dive several days in a row. By the same token, bullet weights may be too long for some female waists: they can be forced up by the hips into the rib cage; it not only hurts whenever you bend over, but it will leave nasty black-and-blue marks on your skin. You might be pleasantly surprised by the comfort of a "soft" belt: the kind that consists of pockets filled with BB's or lead shot.

One solution to the ballast problem is to wear less lead on your belt. Steel tanks are more negatively buoyant in the water than aluminum, and require less compensation. Aluminum tanks can be made heavier by the addition of weights to the clamping strap, or by bolting a lead or steel bar between the connecting bands on doubles. A BC with an integrated weight system may be the answer for you, although it may not supply enough ballast for drysuit diving, and in that regard may have the potential for forcing an uncontrolled ascent if the lead is accidentally jettisoned. On the other hand, you must have enough weight on your belt so that by ditching it you gain adequate positive buoyancy to float you to the surface.

Women are ballasted differently than men. Not only do they have a lower center of gravity, but the extra neoprene that girdles the gluteus adds buoyancy where it is needed the least: fat has inherent positive buoyancy characteristics. (This is not a personal observation, you understand; I'm just passing it along from my advisors.) Change the vertical position of the tank or redistribute your weights in order to achieve a slightly head-up attitude.

Fins made of light plastic instead of heavy rubber will reduce the weight on your feet, thus increasing your leverage, and enable you to deliver a more powerful kick without becoming fatigued.

Lipstick will improve your appearance because it has more ultraviolet protection than lip balm; your lips won't peel as much. Facial cream can also help prevent dry skin by blocking exposure to the sun. For the "raccoon" look, make sure to apply lots of eye make-up before a dive trip. Mascara can help make you appear more bedraggled.

Dive boats have a dress code with the man in mind. Historically, wreck diving has been not only male dominated, but in many cases male exclusive, with the occasional female participant a rarity. When a woman does show up on a boat, her presence can create an awkward situation for all concerned. This is not due to prejudice or a lack of acceptance, but to the physical layout and close confines of the vessel, particularly with regard to the lack of privacy.

Normal social convention is suspended out of necessity when the dive boat cabin becomes a locker room. Donning wetsuits and drysuits requires first removing street clothes, usually down to the skin. In an all male atmosphere no one thinks twice about disrobing in front of others. Throw one woman into the fracas, however, and the climate is as charged as a cyclone in the middle of a Kansas wheat field.

No one is to blame for the disruption in the smooth flow of operations: it's a function of Victorian cultural attitudes reduced by circumstances to absurdity. Nor should anyone fear the adolescent catcalling and locker room high jinks that are relics of college campus enthusiasm. We're all adults here. (Well, most of us are, anyway.) Notwithstanding all of the above, divesting apparel in a mixed sex setting evokes in some people a sense of abject modesty and deep-seated anxiety. If you can't stand the thought of taking off your clothes in front of a roomful of men, it's going to be a hot drive to the shore in a wetsuit or drysuit.

If you don't mind wearing a bathing suit all day, put one on in the morning before you leave home for the dive. If you're diving dry, you can wear your longjohns over it. But if you're going wet, wearing a bathing suit when the day's diving is done is not always an enjoyable option. After an ocean dive, salt will crystallize as the material dries and will irritate the skin. On a chilly day you'll want to put on your street clothes as soon as you get out of your wetsuit; but then your clothes will get wet and you'll be colder than ever. This means that the bathing suit has to come off. One way to do this fashionably is to don an extra large T-shirt and use it as a changing tent.

Not every boat has a "head" (seaman's lingo for water closet) but usually the charter vessels do—and usually only one. You can imagine the pandemonium if a dozen or more people all wanted to get into the head at the same time. Thus the reason why men drop their drawers in the cabin. In this case a woman has to conform to the expedience of the moment—not because of her sex, but because she's an individual in the minority. If she's the only female on board, she can afford the time to change in privacy. But she can't realistically expect every man on the boat to take his turn in line. Under these circumstances, special rules of changing etiquette apply, and the men take over the cabin. You can either step outside for a few minutes, or remain indoors and make a pretense of not looking. Whatever you do, don't complain; it won't do any good, and you'll just wear out your welcome. However, it is the men's responsibility to show a little discretion—deliberate flaunting demands a good scolding.

If you're on a boat without a head, and you've got to change in a room with the guys, make a simple, low-keyed announcement that you would like

In the cold, dark, and turbid water of Lake Huron, Betsy Llewellyn provides scale for the *Regina's* propeller.

them to turn around, and they will generally do as you ask.

Of course, the worst part about diving on a boat without a head is not in having nowhere to change, but in not having anywhere to "go." Center console boats and small charter vessels may not have toilet facilities. That's okay for the guys, who are flexible in most situations, but it's a more delicate procedure for the gals—especially when there is no place on the deck that is not in full public view. Not only must you announce your intentions, but you have to hang your bottom over the gunwale or transom and chance that the waves don't give you a cold water douche. Of course, if the situation calls for more solid action, we're all in the same boat.

What should you do if it's that time of the month? My advisors tell me that they carry on as usual. Unless you are normally prone to debilitating cramps, there's no reason you should change your plans to dive, even during the period of high flow. The stories about attracting sharks is boyish nonsense. And although the changes in hormone level ascribed to premenstral syndrome may increase your sensitivity to nausea and exacerbate seasickness, there is no evidence that the susceptibility to DCS or any other pressure related problems is increased.

What you must know, however, to save yourself embarrassment, is that the plumbing in marine heads is not as tolerant as that in household toilets. The sign that reads, "Do not put anything into the head unless you have eaten it first," is sometimes intended literally, and a separate receptacle is supplied in which used tissue paper is placed. I suggest that you carry a plastic bag for the proper disposal of sanitary products.

A subject that is perhaps of a more sensitive nature is that of married women who dive: not as part of a family outing or vacation, but on their own and without their husbands' accompaniment. Men and women develop different interests as they go through life; so do husbands and wives. The common bonds that hold together a relationship can be many and varied; but this does not imply that all leisure time activity is or should be shared.

Because we live in an intense, frenetic, sometimes all-consuming world—in which a person can easily become lost, like another face in a faceless crowd—people occasionally need to get off on their own, to get away from the pressures of daily routine, to establish a sense of autonomy and individual purpose. In this regard there is no correlation among marriage, motherhood, and wreck diving: they are unconnected aspects of a woman's life.

Common sense dictates that diving during pregnancy represents an unwarranted risk to the unborn child. But as long as a mother's—or father's—family responsibilities are adequately upheld, she has an obligation to herself to achieve her full potential as a person, and to pursue whatever endeavors assume importance in her life. That is her choice.

My good friend Arline Rosenfeld displayed her sense of humor when she allowed us to "load" her onto the boat for this gag shot.

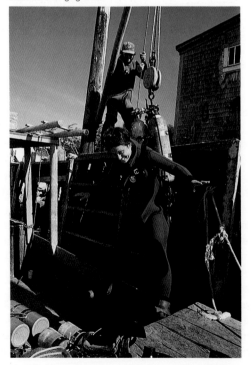

Afterword

Controversy is a necessary evil that stimulates thought.

If tackling the issue of women in diving put my head in a noose, what I write now may serve only to tighten the knot. Yet, my purpose will have been served if my commentary forces people to think, because rational thought is paramount to the stability of modern civilization.

Shipwrecks have become a controversial topic that is constantly in the news. Technology and sophisticated electronics have made it possible to locate and salvage wrecks like never before. Hulls buried in the shallows of rivers and bays can now be detected with ease, deep water exploration is becoming commonplace, and venture capital is readily available to support search and recovery operations at any depth. The rewards of prestige and monetary gain have fostered a new generation of conflict because, inevitably, not everyone wants to share the take.

The "wreck-reational" diver is caught in the middle of a war that marine archaeologists have waged against treasure salvors; they have become the non-combatant casualties shot down by the wild and misplaced fire from the hips of academics who are machine-gunning their way through a sea of shipwrecks so vast that at the present rate of investigation, it will take tens of thousands of years to survey them all—by which time the wrecks will no longer exist. But the machinations of marine archaeologists and their sinister use of political propoganda to further their personal ambitions are beyond the scope of this discussion. The story of the corruption of capitalism in the name of science is a book in itself.

The issues I want to address are: what is the significance of a shipwreck, and how important are its fittings and surviving cargo? In order to do this we must first define the "ship" from which wrecks are derived. Simply stated, a ship is the tractor-trailer of the sea, a truck with propellers instead of eighteen wheels, or a bus that transports people between towns separated by water. Put into perspective, a shipwreck is like a highway accident on a somewhat larger scale: ships are bigger than their land-based counterparts, and may carry more people.

After the passengers and valuables are removed from a bus or a truck or an automobile that is damaged beyond repair, the roadway is cleared so that traffic may resume, and the smashed vehicle is carted off to a junk yard where its parts may be removed and sold. Similarly, a vessel sunk in a shipping lane is salvaged for its restorable assets, then dynamited, wire-dragged, or otherwise removed as a hazard to navigation. What is left of these wrecks—on land as well as under water—is then forsaken and allowed to rot or rust heedlessly. In this context, wreck-diving is equivalent to junk yard exploration, and keepsake collecting is the same as trash picking. How unglamorous.

So where is the controversy, you may ask? And how did marine archaeology get involved. These are valid questions with equally valid answers. The size of a wreck affects only the quantity of the prize, not the quality. What makes the difference between a priceless artifact and a worthless souvenir is age, and the perception of relative value placed upon the object by the finder—and by those who wish to possess it. Riding the tide of historic preservation that has become so popular of late, the bourgeoisie claims that shipwrecks need to be "protected" from the proletariat and preserved for future generations. Do shipwrecks truly need to be preserved? Perhaps. Must *all* shipwrecks be preserved?

Let's change metaphors. Although we maintain national parks and forests, we still fell trees to make lumber and paper products. The fact that we have set aside special woodland conservancies does not mean that *every* tree in the world needs to be spared from timbering, only enough to satisfy other, more esthetic, fancies: a practical evaluation of the matter.

An ancient shipwreck may be a legitimate cultural resource because it can open a window into the lives of our distant ancestors. A modern shipwreck may prove no more enlightening than a demolished Mack truck: a memento of the past with trivial interest. So who makes this distinction between which wrecks are historic and which are not?

Everyone.

Naturally, people's opinions differ. Those who earn their livelihood from shipwrecks and profit from recovering artifacts (marine archaeologists and treasure salvors) view wrecks differently from those who dive for fun and

Nothing remains of the deadeyes on this nineteenth century wreck off Bermuda except for the iron bands that once held them in place. Teredoes have eaten all the wood.

adventure (recreational divers) and who may enjoy picking up trophies of their exploits. But not everyone's opinion has the authority of law. And there's the rub. While the academics would like to legislate their competitors out of business, by raising *every* shipwreck to the status of "historically significant," other user groups take a more reasonable approach and recognize the inherent bias in a system without suitable categories.

For example, wreck-divers of the Great Lakes region have evolved an ethical standard of diving that prohibits the recovery of artifacts, and for a very good reason: the wrecks in the lakes are so well preserved by the cold fresh water that they are in no immediate danger of disintegrating through oxidation or biological decay. Century-old wrecks appear under water today much as they did at the time they went down. Yet these divers recognize that in the salt water environment, where wrecks suffer from devastating collapse and undergo continual deterioration, recovering artifacts is the only way to save them from oblivion.

With this tolerant understanding, all items recovered from the salt water environment are considered to have been "rescued" from the sea, and artifacts are now referred to as "saved material."

Similarly, "significance" is a value judgment, not a quantifible concept.

Mankind could live peacefully if people showed more respect for the values of others: if they tried to gain some introspective insight into modern human nature instead of overemphasizing the importance of past human cultures. Nor can the "wreck war" end until people are willing to accept that there is room in the world for more than one point of view.

These wooden deadeyes, on a nineteenth century wreck in Lake Huron, are nearly perfectly preserved by the cold, fresh water.

BOOKS BY THE AUTHOR

Fiction

Vietnam
Lonely Conflict

Action/Adventure
Mind Set

Supernatural
The Lurking

Science Fiction
Entropy
Return to Mars
Silent Autumn
The Time Dragons Trilogy:
    A Time for Dragons
    Dragons Past
    No Future for Dragons

Nonfiction

Advanced Wreck Diving Guide
Shipwrecks of New Jersey

Track of the Gray Wolf
Wreck Diving Adventures

Available (postage paid) from:

GARY GENTILE PRODUCTIONS
P.O. Box 57137
Philadelphia, PA 19111

Nonfiction

$25 *Andrea Doria: Dive to an Era*
$25 *Ironclad Legacy: Battles of the USS Monitor*
$20 *Primary Wreck-Diving Guide*
$20 *Ultimate Wreck-Diving Guide*
$20 *USS San Diego: the Last Armored Cruiser*
$25 Video (VHS): *The Battle for the USS Monitor*
    The Popular Dive Guide Series:
$20 *Shipwrecks of Delaware and Maryland*
$20 *Shipwrecks of Virginia*
$20 *Shipwrecks of North Carolina: from the Diamond Shoals North*
$20 *Shipwrecks of North Carolina: from Hatteras Inlet South*
Wreck Diving Adventure Novel
$20 *The Peking Papers*